教育部高等学校电子信息类专业教学指导委员会规划教材

高等学校电子信息类专业系列教材·新形态教材

U0168450

通信原理（双语）简明教程

（第2版）

朱艳萍　编著

清华大学出版社

北京

内容简介

本书重点介绍通信系统中各种通信信号的产生、传输和调制解调的基本理论和方法,使学生熟悉并掌握通信系统的基本理论和分析方法,为后续课程打下良好的基础。

在通信系统数学模型和相关数学工具的基础上,本书着重介绍模拟调制系统、数字基带/频带系统、模拟信号的数字传输系统等内容,涉及经典的调制解调、编码译码方法,并注重从系统的角度进行分析和理解。全书共8章,各章之间既独立又相互联系;为了把知识点和相互关系清晰地表示出来,各章首一般都有思维导图。在学习过程中应注重学习方法和技巧的总结,从系统和全局的角度对教材进行整体把握。

本书适合电子信息、通信工程和计算机等相关专业本科生学习使用。

图书在版编目(CIP)数据

通信原理(双语)简明教程/朱艳萍编著.—2版.—北京:清华大学出版社,2024.1
高等学校电子信息类专业系列教材.新形态教材
ISBN 978-7-302-65339-4

Ⅰ.①通… Ⅱ.①朱… Ⅲ.①通信原理-双语教学-高等学校-教材 Ⅳ.①TN911

中国国家版本馆 CIP 数据核字(2024)第 020931 号

责任编辑:古 雪 盛东亮
封面设计:李召霞
责任校对:申晓焕
责任印制:丛怀宇

出版发行:清华大学出版社
    网 址:https://www.tup.com.cn,https://www.wqxuetang.com
    地 址:北京清华大学学研大厦 A 座 邮 编:100084
    社 总 机:010-83470000 邮 购:010-62786544
    投稿与读者服务:010-62776969,c-service@tup.tsinghua.edu.cn
    质量反馈:010-62772015,zhiliang@tup.tsinghua.edu.cn
    课件下载:https://www.tup.com.cn,010-83470236
印 装 者:涿州汇美亿浓印刷有限公司
经 销:全国新华书店
开 本:185mm×260mm 印 张:14.25 字 数:347 千字
版 次:2019 年 7 月第 1 版 2024 年 2 月第 2 版 印 次:2024 年 2 月第 1 次印刷
印 数:1～1500
定 价:69.00 元

产品编号:097173-01

序

FOREWORD

我国电子信息产业占工业总体比重已经超过 10%。电子信息产业在工业经济中的支撑作用凸显，更加促进了信息化和工业化的高层次深度融合。随着移动互联网、云计算、物联网、大数据和石墨烯等新兴产业的爆发式增长，电子信息产业的发展呈现了新的特点，电子信息产业的人才培养面临着新的挑战。

（1）随着控制、通信、人机交互和网络互联等新兴电子信息技术的不断发展，传统工业设备融合了大量最新的电子信息技术，它们一起构成了庞大而复杂的系统，派生出大量新兴的电子信息技术应用需求。这些"系统级"的应用需求，迫切要求具有系统级设计能力的电子信息技术人才。

（2）电子信息系统设备的功能越来越复杂，系统的集成度越来越高。因此，要求未来的设计者应该具备更扎实的理论基础知识和更宽广的专业视野。未来电子信息系统的设计越来越要求软件和硬件的协同规划、协同设计和协同调试。

（3）新兴电子信息技术的发展依赖于半导体产业的不断推动，半导体厂商为设计者提供了越来越丰富的生态资源，系统集成厂商的全方位配合又加速了这种生态资源的进一步完善。半导体厂商和系统集成厂商所建立的这种生态系统，为未来的设计者提供了更加便捷却又必须依赖的设计资源。

教育部 2020 年颁布了新版《高等学校本科专业目录》，将电子信息类专业进行了整合，为各高校建立系统化的人才培养体系，培养具有扎实理论基础和宽广专业技能的、兼顾"基础"和"系统"的高层次电子信息人才给出了指引。

传统的电子信息学科专业课程体系呈现"自底向上"的特点，这种课程体系偏重对底层元器件的分析与设计，较少涉及系统级的集成与设计。近年来，国内很多高校对电子信息类专业课程体系进行了大力度的改革，这些改革顺应时代潮流，从系统集成的角度，更加科学合理地构建了课程体系。

为了进一步提高普通高校电子信息类专业教育与教学质量，推动教育与教学高质量发展，教育部高等学校电子信息类专业教学指导委员会开展了"高等学校电子信息类专业课程体系"的立项研究工作，并启动了"高等学校电子信息类专业系列教材"（教育部高等学校电子信息类专业教学指导委员会规划教材）的建设工作。其目的是推进高等教育内涵式发展，提高教学水平，满足高等学校对电子信息类专业人才培养、教学改革与课程改革的需要。

本系列教材定位于高等学校电子信息类专业的专业课程，适用于电子信息类的电子信息工程、电子科学与技术、通信工程、微电子科学与工程、光电信息科学与工程、信息工程及其相近专业。经过编审委员会与众多高校多次沟通，初步拟定分批次建设约 100 门核心课程教材。本系列教材将力求在保证基础的前提下，突出技术的先进性和科学的前沿性，体现创新教学和工程实践教学；将重视系统集成思想在教学中的体现，鼓励推陈出新，采用"自顶向下"的方法编写教材；将注重反映优秀的教学改革成果，推广优秀的教学经验与理念。

　　为了保证本系列教材的科学性、系统性及编写质量，本系列教材设立顾问委员会及编审委员会。顾问委员会由教指委高级顾问、特约高级顾问和国家级教学名师担任，编审委员会由教育部高等学校电子信息类专业教学指导委员会委员和一线教学名师组成。同时，清华大学出版社为本系列教材配置优秀的编辑团队，力求高水准出版。本系列教材的建设，不仅有众多高校教师参与，也有大量知名的电子信息类企业支持。在此，谨向参与本系列教材策划、组织、编写与出版的广大教师、企业代表及出版人员致以诚挚的感谢，并殷切希望本系列教材在我国高等学校电子信息类专业人才培养与课程体系建设中发挥切实的作用。

吕志伟 教授

前 言

PREFACE

　　为加快电子信息类专业的国际化进程,特编写此双语教程,希望既保证专业课程的正常教学,又兼顾学生专业英语的学习。作者总结了多年来双语教学和英文教学的经验,在内容编写上力争做到深入浅出、循序渐进,并提高教材的可读性和实用性。

　　对于大学本科教学,本书的基本教学时数为 46 学时,同时也能够满足 63 学时的教学。本书在注重基本理论知识的基础上,增加了前后章节知识点的联系和对比,同时增加了通信领域新技术的介绍,注重对专业术语的解释及翻译,能够有效地缩短学生阅读原版英文教程的时间,使学生在有限的学习时间里快速理解和掌握通信技术领域的相关知识,并学以致用。

　　相比于一般的通信原理教材,本书在理论内容基础上,增加了核心章节的实验。首先通过 MATLAB 进行仿真,然后结合实验箱平台进行实际波形的验证。附录涵盖了通信系统中经典调制系统和编码方式的软件仿真与硬件平台实验,能够让学生做到理论联系实际,并提高分析问题和解决问题的能力。

　　此外,本书配套录制了微课视频,为教师教学和学生自学提供参考。

　　在本书的撰写和编辑过程中得到了学生陈梦朝、史涛、巩叙皓等的支持,他们负责完成了不同章节的文字编辑、图表制作等工作,在此向他们表示感谢。再版时,感谢研究生陈家楠、陈继鑫的协助,他们负责完成了课件美化、视频剪辑等工作。特别感谢英国谢菲尔德大学的 Mohammad Reza Anbiyaei 博士为全书英文做了细致的审校。

　　在本书的撰写过程中得到了南京信息工程大学教材建设基金项目、江苏省品牌专业建设工程资助项目、江苏省优秀中青年教师境外研修项目的资助,在此表示感谢。

　　限于作者水平,书中难免有不足之处,恳请读者批评指正。

作　者

2023 年 7 月

教　学　建　议

教学内容	学习要点及教学要求	课 时 安 排	
		全部讲授	部分选讲
第 1 章 绪论	• 了解通信系统发展史。掌握通信系统的模型：基本模型、模拟通信系统模型和数字通信系统模型 • 了解并掌握通信系统的分类 • 掌握衡量通信系统的指标及相应的计算，重点掌握不同进制时，传码率和传信率的计算及二者的关系	2	2
第 2 章 随机过程	• 掌握随机过程的两个不同的定义及数学表示；掌握严平稳和广义平稳随机过程的概念 • 掌握随机过程的数字特征及计算 • 了解高斯随机过程及其性质 • 掌握窄带随机过程的定义及幅度、相位的分布 • 掌握窄带随机过程经过线性系统的数字特征及功率谱密度的变化 • 了解噪声特性，并掌握白噪声的自相关函数和功率谱密度的特性 • 了解正弦波加窄带噪声的幅度、相位分布情况	6～8	6
第 3 章 信道	• 了解并掌握通信系统无线信道和有线信道的分类 • 理解通信编码信道和调制信道的模型，并掌握转移概率的定义及计算 • 重点掌握信息量的定义及计算，掌握连续信道容量即香农定理的公式及计算 • 理解多径效应的原理	2～4	3
第 4 章 连续波 调制系统	• 理解并掌握连续波调制的各种基本概念 • 掌握连续波调制的一般数学模型和系统框图 • 掌握 AM、DSB、SSB、VSB 的调制和解调过程，并能够进行对比 • 理解包络检波和相干解调的不同及优劣 • 掌握调频和调相的基本原理及二者间的关系 • 重点掌握窄带调频和宽带调频的异同，掌握卡森公式，并将 AM 和 NBFM 进行对比 • 理解不同调制系统的调制增益的差异，并能够分析其优劣 • 了解并理解门限效应的定义及现象	7～8	7

续表

教学内容	学习要点及教学要求	课 时 安 排	
		全部讲授	部分选讲
第 5 章 脉冲调制系统	• 了解脉冲调制，掌握几种基本的脉冲调制方式 • 理解模拟信号数字化的三个步骤 • 了解采样的基本原理，掌握采样定理的内容并会应用 • 理解均匀量化和非均匀量化的特点，掌握均匀量化信噪比的计算，掌握非均匀量化的 A 律（A-law）13 折线的量化原理 • 掌握 PCM 编码的原理及计算，了解译码的原理，理解增量调制和 DPCM 的原理	7～9	7
第 6 章 数字基带系统	• 掌握数字基带系统中四种线性码：单极性/双极性归零码、单极性/双极性不归零码的原理及功率谱密度 • 掌握数字基带系统中的传输码（AMI、HDB$_3$ 码）的编译码规律 • 掌握奈奎斯特第一/第二准则的内容及原理 • 了解眼图的产生原理及含义	8	8
第 7 章 数字频带系统	• 掌握数字频带传输系统中的三类基本调制方式（ASK/FSK/PSK）的调制解调原理 • 了解 ASK/FSK/PSK/DPSK 的误码率对比 • 理解 QAM 调制的原理及其星座图 • 理解 OFDM 的基本原理	4～8	4
第 8 章 拓展阅读： 通信系统 新技术	• 理解 UWB 的定义，了解它的几种调制方式，并能够与矩形脉冲调制进行对比 • 了解压缩感知的基本思想及其与传统采样定理的区别，了解压缩感知的发展动态 • 了解 MIMO 系统的优势并熟悉它的应用领域	2～4	2
实验	在理解通信系统的常用调制解调、编码和译码原理基础上，完成以下仿真及实验箱验证实验： • ASK/FSK 调制解调 • AMI/HDB$_3$ 编码和译码 • PCM 编码和译码过程	8～12	8
教学总学时建议		46～63	47

说明：（1）本书为电子信息及相关专业"通信原理"双语课程的教材，理论授课学时数为 46～63 学时，不同专业根据不同的教学要求和计划教学时数可酌情对教材内容进行适当取舍。

（2）本书理论授课学时数中包含习题课、课堂讨论、实验等必要的教学环节。

视 频 目 录

视 频 名 称		时长/分钟	位　　置
第 01 集	通信系统模型与分类	20	第 1 章章首
第 02 集	信息量的计算与通信性能指标	29	1.3 节节首
第 03 集	随机过程的基本概念	33	2.1 节节首
第 04 集	平稳随机过程	24	2.2 节节首
第 05 集	高斯过程与通过一个 LTI 滤波器的随机过程	19	2.3 节节首
第 06 集	窄带高斯过程与正弦波上窄带高斯噪声	26	2.5 节节首
第 07 集	高斯白噪声	14	2.7 节节首
第 08 集	信道的分类	16	3.1 节节首
第 09 集	信道模型及对传输的影响	29	3.2 节节首
第 10 集	信道容量	20	3.4 节节首
第 11 集	调制解调的定义与线性调制	62	4.1 节节首
第 12 集	线性抗噪声性能	25	4.3 节节首
第 13 集	角度调制	49	4.4 节节首
第 14 集	采样过程与 PAM	28	5.1 节节首
第 15 集	量化	29	5.3 节节首
第 16 集	PCM	48	5.4 节节首
第 17 集	波形表示及频谱特性	36	6.1 节节首
第 18 集	匹配滤波	16	6.2 节节首
第 19 集	误码率	12	6.3 节节首
第 20 集	奈奎斯特第一准则	28	6.4.1 节节首
第 21 集	奈奎斯特第二准则	19	6.4.2 节节首
第 22 集	3 种基本数字调制	36	7.1 节节首
第 23 集	抗噪声性能分析	22	7.2 节节首
第 24 集	多进制调制系统(1)	28	7.3 节节首
第 25 集	多进制调制系统(2)	29	7.4 节节首

目 录
CONTENTS

Mind map:

Mind map for

Chapter 1

Analog system: bandwidth efficiency

Digital system: R_b or R_B

Analog system:SNR

Digital system:P_e or P_b

Efficiency

Reliability

WDM

FDM

TDM

CDM

Main performance indexes

Multiplexing

Chapter 1 Introduction

Communication models

The amount of information

Modulation type

General model

Definition

Entropy

Total amount of imfomation

Continuous wave

Pulse

Analog communication model

Digital communication model

Analog modulation

Digital modulation

Analog modulation

Digital modulation

1.1　Basic concepts and models of communication system

1.1.1　Communication system model

There are three basic elements to every communication system：transmitter，channel and receiver，as depicted in Figure 1.1.1.

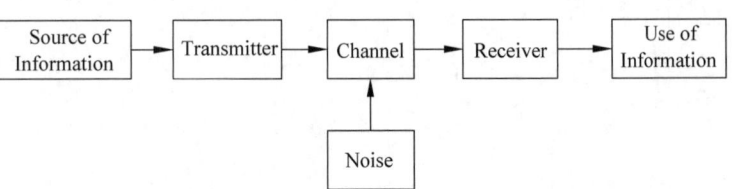

Figure 1.1.1　Elements of a communication system

Source of Information：speech，music，images and computer data are four important sources of information and they may be characterized in terms of the signal that carries the information.

Transmitter：the purpose of the transmitter is to convert the message signal into a form suitable for transmission over the channel.

Channel：the transmitted signal propagates along the channel. In most of the practical cases，interfering signals and also noise are added to the channel output. There are various kinds of channels，such as twisted-pairs（双绞线），coaxial cables（同轴电缆），radio waves，optical fibers（光纤）and etc.

Receiver：has the task of operating on the received signal so as to reconstruct a recognizable form of the original message signal.

The purpose of a communication system is to transmit information—bearing signals through a communication channel.

There are two basic modes of communication：**Broadcasting**（广播、电视等，即一对多）and **Point-to-point communication**（手机、电话等）.

$$\text{Source of Information} \begin{cases} \text{analog signal} \rightarrow \text{speech,music,pictures} \\ \text{digital signal} \rightarrow \text{computer data} \end{cases}$$

1.1.2　Analog communication model

The analog waveform is transmitted in the analog communication system. The analog communication system model is shown in Figure 1.1.2.

Modulation is defined as a process by which some characteristics of a carrier are varied in accordance with a modulating wave.

Demodulation is the reversal of the modulation process. At the receiving end of the system，we usually require the original baseband signal to be recoveried. This is

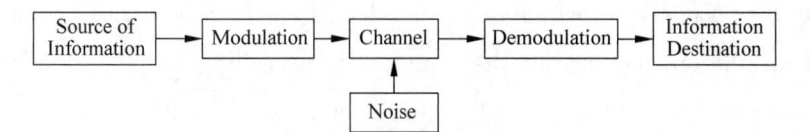

Figure 1.1.2　Analog communication system model

accomplished by demodulation.

　　Baseband signals (基带信号): information-bearing signals which are also refered to as baseband signals.

　　Passband signals (带通信号): the modulated signals (已调信号).

　　Two reasons for studying analog communication are:

　　(1) As long as we hear and see analog communication around us via radio and television, we need to understand that how these communication systems work. Moreover, the study of analog modulation motivates other digital modulation schemes.

　　(2) Analog devices and circuits have a natural affinity for operating at very high speed and they consume very little power compared to their digital counterparts. Accordingly, the implementation of high-speed or very low-power communication systems dictates the use of an analog approach.

1.1.3　Digital communication model

　　In digital communication system, the transmitted information is digital signal. Many digital communication systems can be summarized in the following model as shown in Figure 1.1.3.

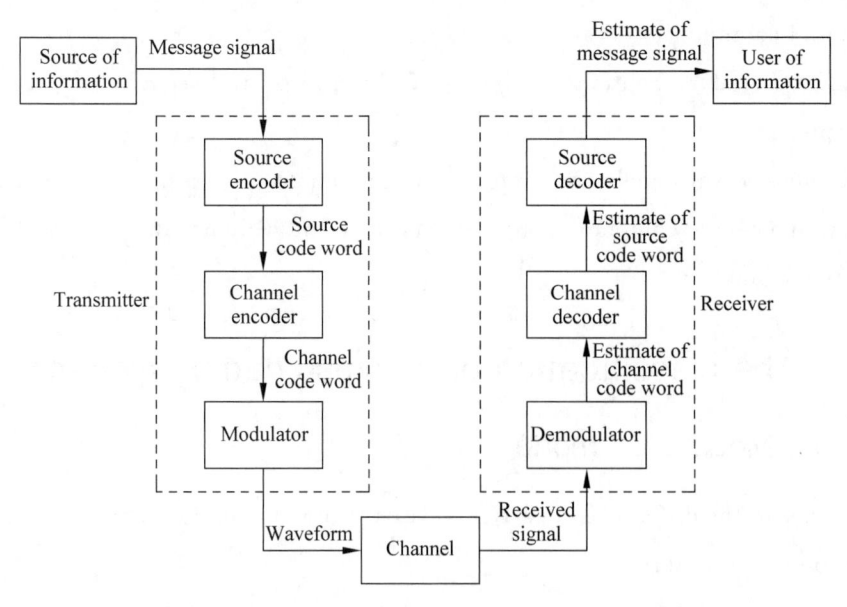

Figure 1.1.3　Digital communication system model

The source encoder removes redundant information from the message signal and is responsible for the efficient use of the channel. Some coding includes the compression, A/D conversion and encryption（加密）coding after compression.

The purpose of the channel coding is to improve the reliability of the signal transmission.

The main purpose of modulation is to make the characteristics of the coded signal adaptive to the channel, and let the modulated signal be successfully transmitted over the channel.

Finally, the common time standard between the transmitter and the receiver is essential in order to know the exact beginning and ending instants of each symbol in the received digital signals. Therefore, there must be a synchronization（同步）circuit in the receiver, which is used to extract the symbol synchronization information from the transmitted signals. Three types of synchronization considered here are as follows: bit synchronization（位同步）, symbol synchronization （码元同步） and code-word synchronization（码字同步）.

Advantages of digital communication:

(1) High anti-interference ability and a decision-making deceiver at the receiver.

(2) Each repeater in the line of retransmissions may reshape the distorted signal, so the accumulation of waveform distortion along the line can be eliminated.

(3) Error correcting techniques, such as error-correcting coding（纠错码）, can be used in digital communication systems.

(4) The digital encryption can be used. Therefore, the security of the system is greatly improved.

(5) Compared with analog communication equipment, the design and manufacture of digital communication equipments are easier, and the volume and the weight of the equipments are smaller.

1.2　The classification of communication systems

1.2.1　Modulation mode

According to the difference of carrier wave, the modulations can be classified different types, as follows in Figure 1.2.1.

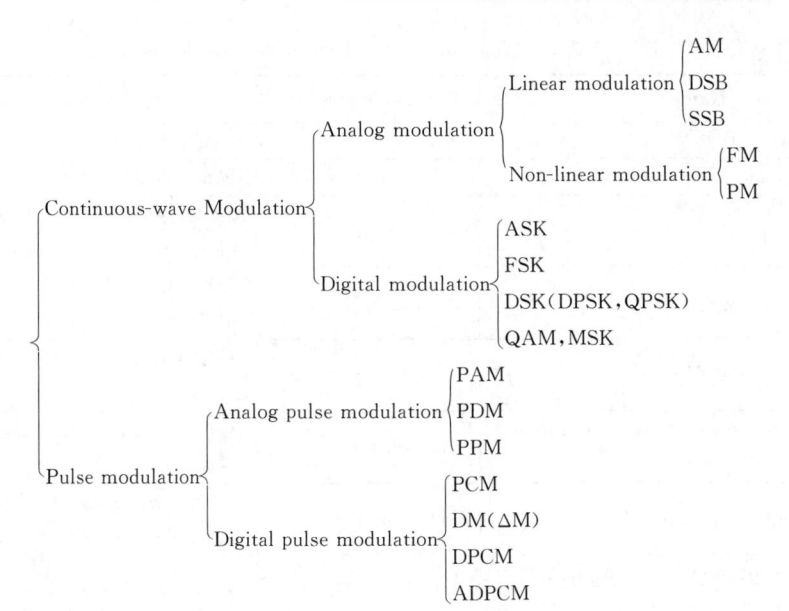

Figure 1.2.1 The common modulation methods

1.2.2 Division of frequency band

In practical applications, the frequency of electromagnetic waves are classified into several frequency bands, which is shown in Table 1.2.1. In addition, the frequency band $1\sim300\mathrm{GHz}$ is usually called microwave band, and it is divided into several bands again as shown in Table 1.2.2.

Table 1.2.1 Modulation methods and applications

Modulation			Application example
Continuous wave modulation	Linear modulation	Conventional bilateral band modulation AM	Broadcasting
		Single-sideband modulation(SSB)	Carrier communication Shortwave radiotelephone communication
		Double-sideband modulation(DSB)	Stereo radio
		Vestigial sideband modulation(VSB)	Television broadcasting Fax
	Nonlinear modulation	Frequency modulation (FM)	Microwave relay Satellite communication Broadcasting
		Phase modulation(PM)	Intermediate modulation method
	Digital modulation	Amplitude shift keying (ASK)	Data transmission
		Frequency shift keying (FSK)	Data transmission
		Phase shift keying(PSK) Differential PSK (DPSK)	Data transmission
		Other efficient digital modulation: Quadrature AM (QAM) Quadrature PSK (QPSK)	Digital microwave Space communication

Table 1.2.2   Frequency band divisions and applications

Name	Frequency range	Applications
Low frequency(LF)	30 to 300kHz	Navigation, time standards
Medium frequency(MF)	300kHz to 3MHz	Marine/aircraft navigation. AM broadcast
High frequency(HF)	3 to 30MHz	AM broadcasting, mobile radio, amateur radio, shortwave broadcasting
Very high frequency(VHF)	30 to 300MHz	Land mobile, FM/TV broadcast, amateur radio
Ultra high frequency(UHF)	300MHz to 3GHz	Cellular phones, mobile radio, wireless LAN, PAN
Super high frequency (SHF), millimeter-wave range	3 to 30GHz	Satellite, radar, backhaul, TV, WLAN, 5G cellular
Extremely high frequency(EHF)	30 to 300GHz	Satellite, radar, backhaul, experimental, 5G cellular
Terahertz, tremendously high frequency (THF) or far infrared(FIR)	300GHz to IR	R&D, experimental

视频讲解

1.3　Information and its measurement

The purpose of communication is to transfer information included in a message. Information is the meaningful or effective content. **Can we find a measure of how much information is produced by a source information**? To answer this question，we note that the idea of information is closely related to that of uncertainty or surprise, but it has no relation to the importance of the message（与事件的不确定性有关，与其重要程度无关）.

The characteristics of the information content：

（1）Information content I contained in a message is a function of the occurrence probability $P(x_k)$ of the message，i. e.

$$I = I[P(x_k)] \qquad (1.3.1)$$

（2）The smaller the occurrence probability of a message，the larger is the information content；and vice versa.

$$\text{When } P(x_k) = 1, \quad I = 0$$
$$\text{When } P(x_k) = 0, \quad I \to \infty$$

（3）Suppose the occurrence probabilities of several independent events are respectively $P(x_1)$, $P(x_2)$, $P(x_3)$, ⋯，then the information content contained in the message consisting of such independent events will equal the sum of information content of the message of each independent event，i. e.

$$I[P(x_1)P(x_2)\cdots] = I[P(x_1)] + I[P(x_2)] + \cdots \qquad (1.3.2)$$

According to the above three properties，the amount of information can be defined as

$$I = \log_a \frac{1}{P(x)} = -\log_a P(x)$$

$$a = 2 \quad \text{unit: bit or b} \tag{1.3.3}$$

Independent equal probability（独立等概时）：

$$M \text{ symbols: } I = \log_2 \frac{1}{1/M} = \log_2 M (\text{b})$$

$$\text{When } M = 2: I = \log_2 \frac{1}{1/2} = \log_2 2 = 1 (\text{b}) \tag{1.3.4}$$

$$\text{or } M = 2^k (k = 1,2,3\cdots): I = \log_2 2^k = k (\text{b})$$

Non-equal probability（不等概时）：

$$\begin{bmatrix} x_1 & x_2 & \cdots & x_M \\ P(x_1) & P(x_2) & \cdots & P(x_M) \end{bmatrix} \text{ and } \sum_{i=1}^{M} P(x_i) = 1 \tag{1.3.5}$$

and their occurrence probability are independent.

If an information source x can generate n different messages: $\{x_1, x_2, \cdots, x_i, \cdots, x_n\}$, then their information contents are: $\{I(x_1), I(x_2), \cdots, I(x_i), \cdots, I(x_n)\}$, therefore, the average information content of the source is defined as:

$$H(x) = E[I(x_k)] = -\sum_{k=1}^{n} P(x_k) \log_2 P(x_k) \tag{1.3.6}$$

$H(x)$ is called as the entropy of the source, which can be regarded as average uncertainty of the source. It is a measure of the average information content per source symbol.

Example 1.3.1: Assume there are 4 possible weather states: clear, cloudy, rainy and foggy. Their occurring probabilities are $1/4, 1/8, 1/8$ and $1/2$. Try to find the entropy of this source.

Solution:

$$H(x) = -\sum_{i=1}^{4} P(x_i) \log_2 P(x_2)$$

$$= -\frac{1}{4} \log_2 \frac{1}{4} - \frac{1}{8} \log_2 \frac{1}{8} - \frac{1}{8} \log_2 \frac{1}{8} - \frac{1}{2} \log_2 \frac{1}{2}$$

$$= \frac{1}{2} + \frac{3}{8} + \frac{3}{8} + \frac{1}{2} = 1.75 (\text{bits/state})$$

1.4　The main performance index of communication systems

The basic factors for measuring the merit of a communication system are effectiveness and reliability（有效性和可靠性）. Efficiency refers to the rate of information transmission in the channel. Reliability refers to the accuracy of the transmission in the channel. The performance index comparisons between analog system and digital system is in Table 1.4.1.

Table 1.4.1　**Main performance index of communication systems**

Performance	Effectiveness	Reliability
Analog system	Transmission frequency/B	Signal to noise ratio (SNR)
Digital system	R_B or R_b	P_b or P_e

1. Analog communication system

Transmission bandwidth: when the message or the signal is transmitted in the channel, the lower is the message bandwidth, the better is the efficiency of the communication system.

SNR (signal to noise ratio) usually is the output SNR of the demodulator at the receiver. The higher SNR, the better is the reliability of the communication system.

2. Digital communication systems

1) Transmission rate

Symbol rate (R_B)（码元传输速率）: the number of symbols transmitted in unit time (s), T is symbols duration and the unit is Baud.

$$R_B = \frac{1}{T} \text{ (Baud)} \tag{1.4.1}$$

Information rate (R_b)（信息传输速率）: the average information content transmitted in unit time, the unit is bit/s(b/s).

For an independent equal probability situation, the relationship between the information rate R_b and the symbol rate R_B is

$$R_b = R_B \log_2 M \text{ (b/s)} \tag{1.4.2}$$

Since the occurrence probability are non-equal, the relationship is

$$R_b = R_B \cdot H(x) \tag{1.4.3}$$

$H(x)$ is the average information content per source symbol.

2) Error probability

Error probability is the main specification of measuring reliability of digital communication. There are two basic definitions.

Symbol error probability P_e: It is the ratio of the number of received symbols in error to the total number of the transmitted symbols, i. e.

$$P_e = \frac{\text{the number of the received \textbf{symbols} in error}}{\text{the total number of the transmitted \textbf{symbols}}} \tag{1.4.4}$$

Bit error probability P_b: It is the ratio of the number of the received bits in error to the total number of the transmitted bits, i. e.

$$P_b = \frac{\text{the number of the received \textbf{bits} in error}}{\text{the total number of the transmitted \textbf{bits}}} \tag{1.4.5}$$

For the binary signals,

$$P_b = P_e$$

Summary and discussion

First, the basic concepts of communication systems are introduced. Then, two main communication models—analog communication and digital communication models are given. Compared with analog communication, digital communication has more superiorities and it is applied widely in nowadays. While, analog communication is the basis, which will be introduced in greater details in Chapter 4. There are lots of connections and comparisons between these two communication systems.

According to the different waveforms of the carrier wave, the communication systems are classified in Figure. 1. 2. 1, which are very important for the entire course.

The definition of the information content is given, and how to calculate it including the entropy of the source should be mastered, which are very helpful in understanding the measurement of the amount of information. Remember, the importance of mathematics in the whole communication systems.

At the end of this chapter, the main performance indices of analogue & digital communication systems are given, especially in Table 1. 4. 1. The definitions and relationship of R_b & R_B(P_e & P_b) should be understood.

This chapter is the basis of communication systems. Try to understand it from mind mapping and have an overview of each chapter from the catalogue.

Homework

1. 1　An information source consists of A, B, C and D. These symbols are represented by codewords 00, 01, 10 and 11. If each binary symbol is transmitted by the pulse with 5ms, then find the average information rates respectively under the following conditions:

(1) The four symbols have equal probability of occurrence(equal probability).

(2) The four symbols have the following probabilities of occurrence.

$$P_A = \frac{1}{5}, \quad P_B = \frac{1}{4}, \quad P_C = \frac{1}{4}, \quad P_D = \frac{3}{10}$$

1. 2　Assume a signal source produces 4-ary signals with equal probability, and the width of symbol is $125\mu s$. Find its symbol rate and information rate.

1. 3　Assume the information rate is 2400b/s of a 4-ary digital transmission system (equal probability). During half an hour, the number of the received symbols in error is 216. Try to find the symbol error probability P_e.

Vocabulary

analog	模拟的	digital	数字的
anti-interference	抗干扰	demodulation	解调
binary	二进制	demodulator	解调器
channel	信道	duration	持续时间
compression	压缩	electromagnetic	电磁的
conversion	转换	encryption	加密

index	指标	receiver	接收机
merit	优点	respectively	分别地
microwave	微波	redundant	冗余的
modulation	调制	specification	规范
occurrence	发生	transmitter	发射机

Mind map:

Mind map for
Chapter 2

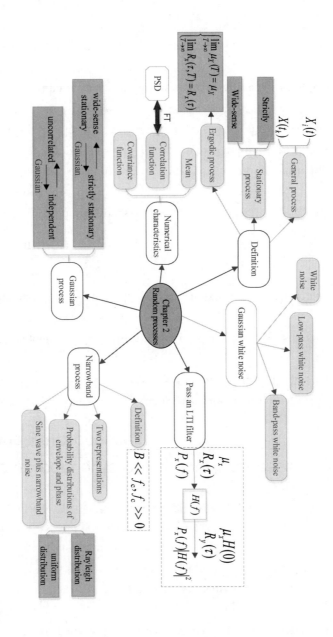

In communication systems，signal and noise are usually random，which are called random signals. Their transmission may be regarded as a random process. Multipath channel and thermal noise are examples of random process. This chapter presents an introductory of **stationary random processes.** In particular，it discusses the following issues：

- The basic concepts of random processes.
- The description of a random process in terms of its mean，correlation and covariance functions.
- Stationary random process and its characteristics，including the ergodicity（各态遍历性）auto-correlation function（自相关函数）and power spectral density（功率谱密度）.
- What happens to a stationary process when it is transmitted through a liner time-invariant filter?
- Narrow band random process and its two representations：
  In-phase and quadrature components（同相分量和正交分量）.
  Envelope and phase components（幅度和相位）.
- Gaussian process and Gaussian white noise.
- Rayleigh and rice distributions（瑞利和莱斯分布）.

视频讲解

2.1 Basic concepts of random processes（definition of random processes）

A mathematical model is widely used to describe a physical phenomenon in physical science and engineering. There are two classes of mathematical models：deterministic and stochastic（random）.

Deterministic signals：The values of the deterministic signals can generally be calculated from a mathematical equation.

Stochastic signals：The values of random signals at any time can not be accurately calculated from a mathematical equation.

2.1.1 Definition

Two properties of random process：
(1) They are functions of time t；
(2) Its value observed at an arbitrary instant is a random variable.

The totality of sample points $s_1, s_2, \cdots, s_n$ corresponding to the aggregate of all possible outcomes of the experiments is called the sample spaces S，as shown in Figure 2.1.1. There are two definitions of a random process.

Definition 1：The sample space or the ensemble composed of sample functions is called a random process $X(t)$.

Random variable（随机变量）$X(t_k)$：From Figure 2.1.1，we note that for a fixed time

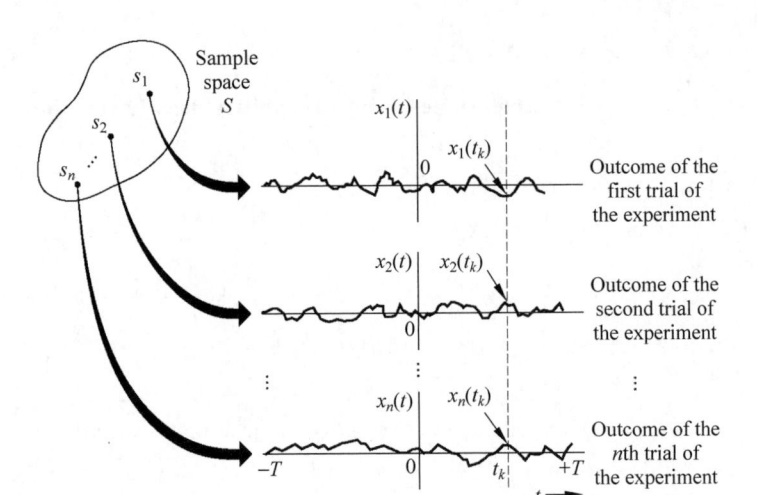

Figure 2.1.1　An ensemble of a sample function

t_k inside the observation interval, the set of numbers $\{x_1(t_k), x_2(t_k), \cdots, x_n(t_k)\}$ constitutes a random variable $X(t_k)$.

Definition2: An indexed ensemble (family) of random variable $X(t)$ is called a random process.

Distinguish between $X(t_k)$ and $X(t)$:

For a random variable, the outcome of a random experiment is mapped into a number.

For a random process, the outcome of a random experiment is mapped into a waveform that is a function of time.

2.1.2　Numerical characteristics of a random process

Under most conditions, the probability distribution of a random process is very difficult to determine by experimental method.

Mean, variance and correlation function can partially describe the statistical characteristics of a random process, which are often used in the research of communication systems. Their definitions and basic properties are described respectively as follows:

1. Mean (mathematic expectation: 数学期望)$\mu_X(t)$

The statistical mean of a random process $X(t)$ is defined by

$$\mu_X(t) = E[X(t)] = \int_{-\infty}^{\infty} x f_x(x,t) \, \mathrm{d}x \tag{2.1.1}$$

where $f_x(x,t)$ is the first-order probability density function of the process.

2. Variance

Define the variance of a random process $X(t)$ as:

$$\delta_x^2(t) = D[x(t)] = E\{[x(t) - \mu_X(t)]^2\} \tag{2.1.2}$$

It can describe the deviation at the instant t from its mathematical expectation.

3. Correlation function (相关函数)$R(t_1, t_2)$ and covariance function (协方差函数) $B(t_1, t_2)$

Autocorrelation function(自相关函数):

$$R_x(t_1,t_2) = E[X(t_1)X(t_2)] = \int_{-\infty}^{\infty}\int_{-\infty}^{\infty} x_1 x_2 f(x_1,x_2;t_1,t_2)\,\mathrm{d}x_1\mathrm{d}x_2 \quad (2.1.3)$$

where $f(x_1,x_2;t_1,t_2)$ is a **2-dimensioned joint probability density function**（二维联合概率密度函数）.

Covariance function（协方差函数）：

$$B_x(t_1,t_2) = E\{[X(t_1)-\mu_X(t_1)][X(t_2)-\mu_X(t_2)]\}$$

$$= \int_{-\infty}^{\infty}\int_{-\infty}^{\infty}[x(t_1)-\mu_X(t_1)][x(t_2)-\mu_X(t_2)]f(x_1,x_2;t_1,t_2)\mathrm{d}x_1\mathrm{d}x_2 \quad (2.1.4)$$

The relationship between $R(t_1,t_2)$ and $B(t_1,t_2)$ is

$$B(t_1,t_2) = R(t_1,t_2) - \mu(t_1)\mu(t_2) \quad (2.1.5)$$

$R(t_1,t_2)$ and $B(t_1,t_2)$ express the correlation degree of two random values obtained by sampling one random process at two instants.

For two random processes $X(t), Y(t)$, the corresponding $R(t_1,t_2)$ and $B(t_1,t_2)$ are:

Cross-correlation function（互相关函数）

$$R_{XY}(t_1,t_2) = E[X(t_1)Y(t_2)] \quad (2.1.6)$$

Cross covariance function（互协方差函数）

$$B_{XY}(t_1,t_2) = E[X(t_1)-\mu_X(t_1)][Y(t_2)-\mu_Y(t_2)] \quad (2.1.7)$$

2.2 Stationary random process

视频讲解

Stationary random process is a widely used process, which plays an important role in communication systems.

2.2.1 Definition

Definition 1：Stationary in strict-sense（严平稳）

If the statistic characteristic of a random process is independent of the time origin, then the random process is called a stationary process in its strict meaning or a strict stationary random. Its mathematical definition is：

$$f(x_1,x_2,\cdots,x_n;t_1,t_2,\cdots,t_n) = f(x_1,x_2,\cdots,x_n;t_1+\Delta,t_2+\Delta,\cdots,t_n+\Delta)$$

$$(2.2.1)$$

The (1-dimensional distribution function) is independent of time t：

$$f(x_1,t_1) = f(x_1) \quad (2.2.2)$$

The (2-dimensional distribution function) is related only to the time interval τ between t_1 and t_2

$$f(x_1,x_2;t_1,t_2) = f(x_1,x_2;\tau) \quad (2.2.3)$$

Definition 2：Stationary in wide-sense（广义平稳）

Consider a strictly stationary random process $X(t)$. The mean of the process $X(t)$ is

$$\mu_X(t) = E[X(t_1)] = \int_{-\infty}^{\infty} x f(x)\mathrm{d}x = \mu_X, \quad \text{for all } t \quad (2.2.4)$$

$f(x)$ is independent of time t.

The mean of a strictly stationary process is a constant.

The autocorrelation function of the process $X(t)$ is written as:

$$R_x(t_1,t_2)=E[X(t_1)X(t_2)]=\int_{-\infty}^{\infty}\int_{-\infty}^{\infty}x_1x_2 f(x_1,x_2;\ t_1,t_2)\,\mathrm{d}x_1\mathrm{d}x_2$$

$$=\int_{-\infty}^{\infty}\int_{-\infty}^{\infty}x_1x_2 f(x_1,x_2;\ \tau)\,\mathrm{d}x_1\mathrm{d}x_2=R_x(\tau) \tag{2.2.5}$$

The autocorrelation of a strictly stationary process depends only on the time difference $\tau(\tau=t_2-t_1)$.

$$\begin{cases}E[X(t)]=\mu_X=\text{constant}\\ R_x[t_1,t_2]=R_x(t_2-t_1)=R(\tau)\end{cases} \tag{2.2.6}$$

The class of random process that satisfy the above equations are called wide-sense stationary random process.

The relationship between strict-sense stationary and wide-sense stationary is

$$\boxed{\text{strict-sense stationary}\ \underset{\times}{\overset{\surd}{\longleftrightarrow}}\ \text{wide-sense stationary}}$$

A wide-sense random process is not necessarily strictly stationary.

In communication system theory, it is often regarded as that random signal and noise are wide-sense stationary random processes.

2.2.2　Ergodicity

In practice, it is impossible to calculate the statistical mean or autocorrelation for all realizations of the random processes. However, if a random process has ergodicity, then its statistic mean is equal to its time average. Hence, the statistical mean may be replaced by the time average of an arbitrary realization of an ergodic process.

Consider the sample function of a stationary process $X(t)$, with the observation interval defined as $-T\leqslant t\leqslant T$. The time average and time auto-correlation are defined as:

$$\begin{cases}\mu_X(T)=\lim_{T\to\infty}\dfrac{1}{T}\int_{-\frac{T}{2}}^{\frac{T}{2}}x(t)\,\mathrm{d}t\\ R_x(\tau,T)=\lim_{T\to\infty}\dfrac{1}{T}\int_{-\frac{T}{2}}^{\frac{T}{2}}x(t)x(t+\tau)\,\mathrm{d}t\end{cases} \tag{2.2.7}$$

If the following two conditions are satisfied:

$$\begin{cases}\mu_X(T)=\mu_X\\ R_x(\tau,T)=R_x(\tau)\end{cases} \tag{2.2.8}$$

The process is called ergodic. According to this property, it is not necessary to make infinite observations, but it is only necessary to make one observation, and use time average instead of statistic mean, then the calculation can be reduced tremendously.

If a random process has ergodicity, then it must be a strict stationary random process. However, a strict stationary random process is not always ergodic.

$$\boxed{\text{ergodic random process}\ \underset{\times}{\overset{\surd}{\longleftrightarrow}}\ \text{strict stationary random process}}$$

In practice, when we analyze the stationary state of most communication systems, it is always assumed that means and autocorrelation functions of the signal and noise are all ergodic.

Example 2.2.1: Consider a sinusoidal signal with random phase, defined by

$$X(t) = A\cos(2\pi f_c t + \theta) \tag{2.2.9}$$

where A and f_c are constant and θ is a random variable that is uniformly distributed over the interval $[0, 2\pi]$, that is

$$f(\theta) = \begin{cases} \dfrac{1}{2\pi}, & 0 \leqslant \theta \leqslant 2\pi \\ 0, & \text{elsewhere} \end{cases} \tag{2.2.10}$$

Discuss whether $X(t)$ is ergodicity.

Answer:

(1) The statistic average is

$$
\begin{aligned}
\mu_X(t) &= E[X(t)] \\
&= \int_0^{2\pi} A\cos(2\pi f_c t + \theta) \cdot \frac{1}{2\pi} d\theta \\
&= \frac{A}{2\pi} \int_0^{2\pi} (\cos\omega_c t \cos\theta - \sin\omega_c t \sin\theta) d\theta \\
&= \frac{A}{2\pi} \left[\cos\omega_c t \int_0^{2\pi} \cos\theta\, d\theta - \sin\omega_c t \int_0^{2\pi} \sin\theta\, d\theta \right] \\
&= 0
\end{aligned}
$$

Autocorrelation function is

$$
\begin{aligned}
R(t_1, t_2) &= E[x(t_1) x(t_2)] \\
&= E[A\cos(\omega_c t_1 + \theta) \cdot A\cos(\omega_c t_2 + \theta)] \\
&= \frac{A^2}{2} E\{\cos\omega_c(t_2 - t_1) + \cos[\omega_c(t_2 + t_1) + 2\theta]\} \\
&= \frac{A^2}{2} \cos\omega_c(t_2 - t_1) + \frac{A^2}{2} \int_0^{2\pi} \cos[\omega_c(t_2 + t_1) + 2\theta] \cdot \frac{1}{2\pi} d\theta \\
&= \frac{A^2}{2} \cos\omega_c(t_2 - t_1) + 0 \\
&= \frac{A^2}{2} \cos\omega_c \tau
\end{aligned}
$$

Hence, $x(t)$ is a wide-sense stationary process.

The autocorrelation function is plotted in Figure 2.2.1.

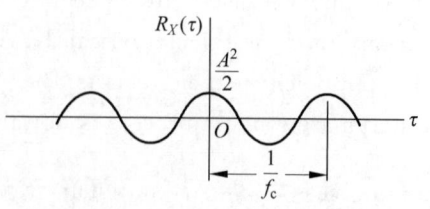

Figure 2.2.1　Autocorrelation function of a sine wave with random phase

(2) The time average is

$$\mu_X(T) = \lim_{T \to \infty} \frac{1}{T} \int_{-\frac{T}{2}}^{\frac{T}{2}} A\cos(\omega_c t + \theta)\,\mathrm{d}t = 0 = \mu_X(t)$$

$$R_X(\tau, T) = \lim_{T \to \infty} \frac{1}{T} \int_{-\frac{T}{2}}^{\frac{T}{2}} A\cos(\omega_c t + \theta) \cdot A\cos[\omega_c(t+\tau) + \theta]\,\mathrm{d}t$$

$$= \lim_{T \to \infty} \frac{A^2}{2T} \left\{ \int_{-\frac{T}{2}}^{\frac{T}{2}} \cos\omega_c \tau\,\mathrm{d}t + \int_{-\frac{T}{2}}^{\frac{T}{2}} \cos(2\omega_c t + \omega_c \tau + 2\theta)\,\mathrm{d}t \right\}$$

$$= \frac{A^2}{2} \cos\omega_c \tau$$

Since $\mu_X(t) = \mu_X(T)$, $R_X(\tau) = R_X(\tau, T)$, so $X(t)$ is an ergodic process.

2.2.3   Autocorrelation function of stationary random processes

Let $X(t)$ be a stationary random process, we redefine the autocorrelation function of $X(t)$ as

$$R_X(\tau) = E[X(t+\tau)X(t)] \quad \text{(for all } t)$$

This autocorrelation function has five important properties:

(1) $R(0) = E[X^2(t)] = P_X$ —The average power of $X(t)$

(2) $R(\tau) = R(-\tau)$ —An even function of τ

(3) $|R(\tau)| \leqslant R(0)$

According to

$$E[X(t) \pm X(t+\tau)]^2 \geqslant 0$$

We can get

$$E[X^2(t) + X^2(t+\tau) \pm 2X(t)X(t+\tau)] \geqslant 0$$
$$2R(0) \pm 2R(\tau) \geqslant 0$$
$$|R(\tau)| \leqslant R(0)$$

And $R(\tau)$ has its maximum magnitude at $\tau = 0$.

(4) $R(\infty) = E^2[X(t)] = \mu^2$ —The power of the D. C. component (直流功率)

$$\lim_{t \to \infty} R(\tau) = \lim_{t \to \infty} E[X(t+\tau)X(t)]$$
$$= E[X(t)]E[X(t+\tau)]$$
$$= E^2[X(t)]$$

Therefore, when $\tau \to \infty$, $X(t)$ and $X(t+\tau)$ are statistically independent of each other.

(5) $R(0) - R(\infty) = \delta_X^2$ —A. C. power (交流功率)

When $\mu_X = 0$, $R(\infty) = E^2[X(t)] = 0$, the result is $R(0) = \delta_X^2$.

2.2.4   Power spectral density

Let us discuss the power spectral density (PSD) $P_X(f)$ of a stationary random process $X(t)$. We know that the PSD of a deterministic power signal $X(t)$ may be expressed as:

$$P(f) = \lim_{T \to \infty} \frac{|S_T(f)|^2}{T} \tag{2.2.11}$$

$S_T(f)$ is the frequency spectrum function of the truncated function （截短函数） of $S(f)$ (see in Figure 2. 2. 2).

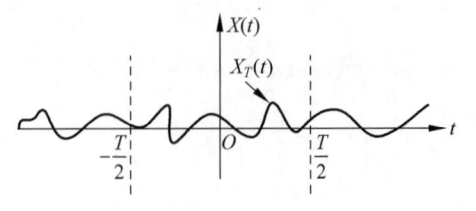

Figure 2. 2. 2　Truncated function

Similarly，the PSD of a random process can be expressed or defined by equation (2. 2. 11).

Let the PSD of a stationary random process $X(t)$ be $P_X(f)$, the truncated function of a realization of $X(t)$ and $X_T(t)$, and the Fourier transform of $X_T(t)$ be $S_T(f)$, then we have

$$P_X(f) = E[p(f)] = \lim_{T \to \infty} \frac{E|S_T(f)|^2}{T} \qquad (2. 2. 12)$$

This is the expression of the PSD of a stationary random process.

The power spectral density $P_X(f)$ and the autocorrelation function $R_X(\tau)$ of a stationary process $X(t)$ form a Fourier-transform pair with τ and f as the variables of interest，as shown by the pair of relations

$$P_X(f) = \int_{-\infty}^{\infty} R_X(\tau) \exp(-j2\pi f\tau) d\tau \qquad (2. 2. 13)$$

$$R_X(\tau) = \int_{-\infty}^{\infty} P_X(f) \exp(j2\pi f\tau) df \qquad (2. 2. 14)$$

Equations (2. 2. 13) and (2. 2. 14) constitute the Einstein-Wiener-Khintchine relations. The relations show that if either $R_X(\tau)$ or $P_X(f)$ of a random process is known，the other can be found exactly.

Using this pair of relations，we can derive some general properties：

(1) The total power of a stationary process is：$R(0) = \int_{-\infty}^{\infty} P_X(f) df$.

(2) $P_X(f) \geqslant 0$　　for all f　—$P_X(f)$ is always nonnegative.

(3) $P_X(-f) = P_X(f)$　　　—$P_X(f)$ is an even function （偶函数）.

Example 2. 2. 2：Find the autocorrelation and the PSD of sinusoidal wave with random phase

$$X(t) = A\cos(2\pi f_c\tau + \theta)$$

Answer：

From example 2. 2. 1，we have calculated the autocorrelation function

$$R_X(\tau) = \frac{A^2}{2}\cos(2\pi f_c\tau) = \frac{A^2}{2}\cos(\omega_c\tau)$$

Taking the Fourier transform of both sides of the relation defining $R_X(\tau)$, we find that PSD of the sinusoidal process $X(t)$ is

$$P_X(f) = \frac{A^2}{4}[\delta(f - f_c) + \delta(f + f_c)]$$

$$P_X(\omega) = \frac{A^2}{2}\pi[\delta(\omega - \omega_c) + \delta(\omega + \omega_c)]$$

which consists of a pair of δ function weighed by the factor $\frac{A^2}{4}$ and located at $\pm f_c$. The average power is：

$$S = R(0) = \frac{A^2}{2}\cos(\omega_c \cdot 0) = \frac{A^2}{2}$$

2.3　Gaussian process

视频讲解

In this section，we consider an important random process known as Gaussian process.

1. Definition of the Gaussian process

The Gaussian Process is also called normal random process. The thermal noise in communication systems is usually a Gaussian process.

The one dimensional probability density function of the Gaussian process $X(t)$ obeys the normal distribution. It can be expressed as：

$$f(x) = \frac{1}{\sqrt{2\pi}\delta} \cdot \exp\left[-\frac{(x-a)^2}{2\delta^2}\right] \tag{2.3.1}$$

The definition of the Gaussian process $X(t)$ with the joint probability density function（联合概率密度函数）of n dimension satisfies the following condition：

$$f_n(x_1, x_2, \cdots, x_n; t_1, t_2, \cdots, t_n) =$$

$$\frac{1}{(2\pi)^{\frac{n}{2}}\delta_1\delta_2\cdots\delta_n|\boldsymbol{B}|^{\frac{1}{2}}}\exp\left[\frac{-1}{2\boldsymbol{B}}\sum_{j=1}^{n}\sum_{k=1}^{n}|\boldsymbol{B}|_{jk}\left(\frac{x_j - a_j}{\delta_j} \cdot \frac{x_k - a_k}{\delta_k}\right)\right] \tag{2.3.2}$$

where a_k is the mathematical expectation（statistical mean）of x_k, δ_k is the standard deviation of x_k and $|\boldsymbol{B}|$ is the determinant of the normalized covariance matrix：

$$|\boldsymbol{B}| = \begin{vmatrix} 1 & b_{12} & \cdots & b_{1n} \\ b_{21} & 1 & \cdots & b_{2n} \\ \vdots & \vdots & & \vdots \\ b_{n1} & b_{n2} & \cdots & 1 \end{vmatrix}$$

$|\boldsymbol{B}|_{jk}$ is the algebraic cofactor（代数余子式）of the element b_{jk} in the determinant $|\boldsymbol{B}|$; b_{jk} is the normalized covariance function , i. e.

$$b_{jk} = \frac{E\left[|(x_j - a_j)(x_k - a_k)|\right]}{\sigma_j\sigma_k}$$

This probability density is decided only by each mathematical expectation，standard deviation and normalized covariance of the random variable，so it is a wide-sense stationary random process.

N-dimensional joint probability density function can also be defined in matrix form as

$$f_n(x_1,x_2,\cdots,x_n; t_1,t_2,\cdots,t_n) = \frac{1}{(2\pi)^{\frac{n}{2}}|\boldsymbol{B}|^{\frac{1}{2}}} \exp\left[-\frac{1}{2}(\boldsymbol{x}-\boldsymbol{\mu})^{\mathrm{T}}\boldsymbol{\Sigma}^{-1}(\boldsymbol{x}-\boldsymbol{\mu})\right]$$

where the superscript（上标）T denotes transposition（转置）and

$\boldsymbol{\mu} = $ mean vector $= [\mu_1,\mu_2,\cdots,\mu_n]^{\mathrm{T}}$

$\boldsymbol{\Sigma} = $ covariance matrix $= \{C_x(t_k,t_i)\}^n \quad k,i=1,2,\cdots,n$

$C_x(t_k,t_i) = E\{[X(t_k)-\mu_X(t_k)][X(t_i)-\mu_X(t_i)]\} \quad k,i=1,2,\cdots,n$

$\boldsymbol{\Sigma}^{-1} = $ inverse of covariance matrix

$\boldsymbol{B} = $ determinant of covariance matrix

2. Some useful properties of Gaussian process

If a Gaussian process is stationary, then process is also strict-sense stationary.

$$\text{wide-sense stationary} \underset{\text{Gaussian }\surd}{\overset{\surd}{\longleftrightarrow}} \text{strict-sense stationary}$$

If $x_1,x_2,\cdots,x_n$ are uncorrelated, that is

$$E\{[X(t_k)-\mu_X(t_k)][X(t_i)-\mu_X(t_i)]\}=0, \quad i\neq k \quad \text{or} \quad b_{jk}=0$$

Then, equation(2.3.2) is reduced to

$$f_n(x_1,x_2,\cdots,x_n; t_1,t_2,\cdots,t_n) = \prod_{k=1}^{n} \exp\left[-\frac{(x_k-a_k)^2}{2\sigma_k^2}\right] = f(x_1,t_1)f(x_2,t_2)\cdots f(x_n,t_n)$$

The above equation shows that these random variables are statistically independent（统计独立）.

Two uncorrelated random variables are not always independent of each other, and the two independent random variables are certainly uncorrelated.

$$\text{uncorrelated} \underset{\surd}{\overset{\times}{\longleftrightarrow}} \text{independent}$$
$$\text{uncorrelated} \underset{\surd}{\overset{\text{Gaussian}\surd}{\longleftrightarrow}} \text{independent}$$

3. Gaussian random variable

The probability density of random distribution curve is shown in Figure 2.3.1, and this distribution has the following properties.

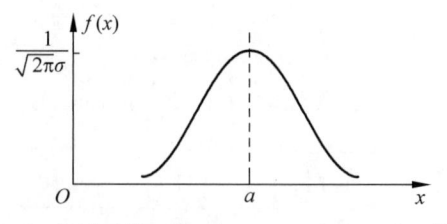

Figure 2.3.1   Probability density curve of Gaussian distribution

(1) $f(x)$ is symmetrical to the straight line $x=a$, $f(a+x)=f(a-x)$.

(2) $\displaystyle\int_{-\infty}^{\infty} f(x)\mathrm{d}x=1$ and $\displaystyle\int_{-\infty}^{a} f(x)\mathrm{d}x=\int_{a}^{+\infty} f(x)\mathrm{d}x=\frac{1}{2}$.

(3) if $a=0$, $\sigma=1$, then the distribution is called the standardized normal distribution.

$$f(x) = \frac{1}{2\pi} \exp\left(-\frac{x^2}{2}\right)$$

The normal distribution function can be expressed as:

$$F(x) = \int_{-\infty}^{x} \frac{1}{\sqrt{2\pi}\delta} \exp\left[-\frac{(z-a)^2}{2\delta^2}\right] dz = \Phi\left(\frac{x-a}{\delta}\right)$$

where $\Phi(x)$ is called the probability integral function, its definition is

$$\Phi(x) = \frac{1}{\sqrt{2\pi}} \int_{-\infty}^{x} \exp\left[-\frac{z^2}{2}\right] dz$$

This integral is difficult to calculate. Approximation of the integral for different values of x can usually be found by lookup-table method.

The following function describes the definition of error:

$$\text{erf}(x) = \frac{2}{\sqrt{\pi}} \int_{0}^{x} e^{-z^2} dz$$

Lookup-table method (查表方法) is also usually used to obtain the error function $\text{erf}(x)$.

The complementary error function can be approximated as

$$\text{erfc}(x) \approx \frac{1}{x\sqrt{\pi}} e^{-x^2}$$

When we discuss anti-noise performance of communication system later, the above error function $\text{erf}(x)$ and complementary error function $\text{erfc}(x)$ are often used. The table of error function is given in Appendix C.

2.4　Transmission of a random process through a linear time-invariant (LTI) filter

Suppose that a random process $X(t)$ is applied as input to a linear time-invariant (LTI: 线性时不变) filter with impulse response $h(t)$, producing a new random process $Y(t)$ at the filter output. The linear system sketch is drawed in Figure 2.4.1.

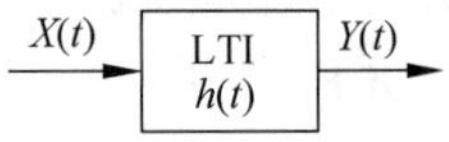

Figure 2.4.1　Linear system

The output random process $Y(t)$ can be written as

$$Y(t) = \int_{0}^{\infty} h(\tau) X(t-\tau) d\tau \tag{2.4.1}$$

The transmission of a process through a linear time-invariant filter is governed by the convolution.

Assume $X(t)$ is a stationary random process, let us analyze the statistic characteristics of the output random process $Y(t)$. First, we find the numerical characteristics and PSD of the output $Y(t)$, and then we discuss its probability distribution.

1. Mathematical expectation $E[Y(t)]$

The mean of $Y(t)$ is

$$\mu_Y(t) = E[Y(t)]$$

$$= E\left[\int_{-\infty}^{\infty} h(\tau)X(t-\tau)\mathrm{d}\tau\right]$$

$$= \int_{-\infty}^{\infty} h(\tau)E[X(t-\tau)]\mathrm{d}\tau \qquad (2.4.2)$$

Since the input random process $X(t)$ is stationary, So the mean $\mu_X(\tau)$ is a constant μ_X and the equation(2.4.2) may be simplified as follows:

$$\mu_Y = \mu_X \int_{-\infty}^{\infty} h(\tau)\mathrm{d}\tau$$

$$= \mu_X \cdot H(0)$$

where $H(0)$ is the zero-frequency(DC) response of the system.

We have

$$E[Y(t)] = \mu_Y \cdot H(0) = \text{constant}$$

2. Autocorrelation function $R_Y(t_1, t_1 + \tau)$ of $Y(t)$

According to the definition of the autocorrelation function, we have

$$R_Y(t_1, t_1 + \tau) = E[Y(t_1)Y(t_1 + \tau)]$$

$$= E\left[\int_{-\infty}^{\infty} h(u)X(t_1 - u)\mathrm{d}u \int_{-\infty}^{\infty} h(v)X(t_1 + \tau - v)\mathrm{d}v\right]$$

$$= \int_{-\infty}^{\infty}\int_{-\infty}^{\infty} h(u)h(v)E[X(t_1 - u)X(t_1 + \tau - v)]\mathrm{d}u\,\mathrm{d}v$$

The input $X(t)$ is a stationary random process:

$$E[X(t_1 - u)X(t_1 + \tau - v)] = R_X(\tau + u - v)$$

then

$$R_Y(t_1, t_1 + \tau) = \int_{-\infty}^{\infty}\int_{-\infty}^{\infty} h(u)h(v)E[X(t_1 - u)X(t_1 + \tau - v)]\mathrm{d}u\,\mathrm{d}v = R_Y(\tau)$$

$Y(t)$ is only related to the time interval τ, and is independent of the time t_1.

On combining this result with that involving the mean μ_Y, we see that **if the input to a LTI filter is a stationary process, then the output is also a stationary process.**

3. Power spectral density $P_Y(f)$ of $Y(t)$

The $R(\tau)$ and $P(f)$ of a random process are a Fourier transform pair, so

$$P_Y(f) = \int_{-\infty}^{\infty} R_Y(\tau)\mathrm{e}^{-\mathrm{j}\omega\tau}\mathrm{d}\tau$$

$$= \int_{-\infty}^{\infty} \mathrm{d}\tau \int_{-\infty}^{\infty} \mathrm{d}u \int_{-\infty}^{\infty} h(v)h(v)R_X(\tau + u - v)\mathrm{e}^{-\mathrm{j}\omega\tau}\mathrm{d}v$$

Let $\tau' = \tau + u - v$, by substituting it into the above equation, we have

$$P_Y(f) = \int_{-\infty}^{\infty} h(u)\mathrm{e}^{\mathrm{j}\omega u}\mathrm{d}u \int_{-\infty}^{\infty} h(v)\mathrm{e}^{-\mathrm{j}\omega v}\mathrm{d}v \int_{-\infty}^{\infty} R_X(\tau')\mathrm{e}^{-\mathrm{j}\omega\tau'}\mathrm{d}\tau'$$

$$= H^*(f)H(f)P_X(f) = |H(f)|^2 P_X(f)$$

The result shows that the PSD of $Y(t)$ is equal to the PSD of $X(t)$ multiplied by

$|H(f)|^2$. Both input-output time-domain and frequency-domain changes are clearly shown in Figure 2.4.2.

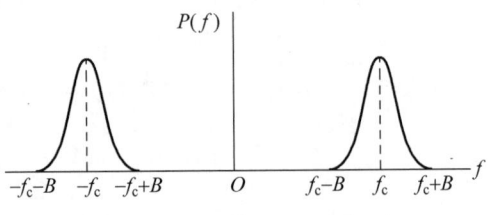

Figure 2.4.2   The input-output relation of an LTI system in time-domain and frequency-domain

4. Probability distribution of $Y(t)$

If the input is a random Gaussian process，then the output is also a random Gaussian process（如果线性系统输入过程是高斯的，则系统的输出过程也是高斯的）.

This principle is based on the following limit of a summation：

$$Y(t) = \lim_{\tau \to \infty} \sum_{k=0}^{\infty} X(t - \tau_k) h(\tau_k) \Delta\tau_k$$

$X(t)$ is a Gaussian process，therefore each term $X(t - \tau_k) h(\tau_k) \Delta\tau_k$ is a normal random variable at an arbitrary instant. The summation of infinite normal random variables is also a normal random variable. Therefore，the output random process is also a normal random process. The difference between input and output random is the numerical characteristics.

2.5　Narrowband random process

视频讲解

In communication systems，signals and noises are often limited to the narrow band，especially the noise. Suppose the bandwidth of the random process is Δf , and the central frequency is f_c.

If $\Delta f \ll f_c$ and $f_c \gg 0$，then the random process is a narrowband random process（窄带随机过程）, as shown in Figure 2.5.1.

Figure 2.5.1   PSD of a narrow band process

The waveform of a narrowband process looks like a sinusoidal wave with slowly varying envelop and phase，as shown in Figure 2.5.2. It can be expressed as：

$$X(t) = a_x(t) \cos[\omega_c t + \varphi_x(t)], \quad a_x(t) \geqslant 0 \qquad (2.5.1)$$

where $a_x(t)$ and $\varphi_x(t)$ are random envelope and random phase of $X(t)$. Obviously，the variation of $a_x(t)$ and $\varphi_x(t)$ is slower than the variation of the carrier $\cos\omega_c t$.

This equation is defined in terms of two components called envelope and phase.

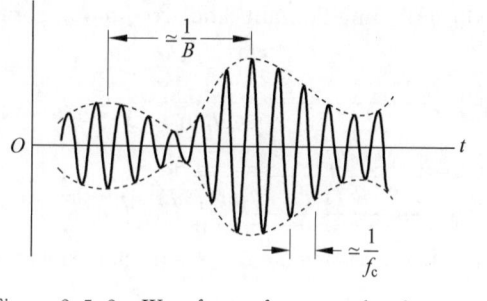

Figure 2.5.2 Waveform of a narrowband process

From equation(2.5.1)，we can deduce another mathematical representation，which is called in-phase and quadrature components.

$$X(t) = X_c(t)\cos\omega_c t - X_s(t)\sin\omega_c t \qquad (2.5.2)$$

where $X_c(t) = a_x(t)\cos\varphi_x(t)$　(in-phase components)

$X_s(t) = a_x(t)\sin\varphi_x(t)$　(quadrature components)

From these two representatives，the statistic characteristics of $X(t)$ can be determined by the statistic characteristic of $a_x(t)$ and $\varphi_x(t)$ or $X_c(t)$ and $X_s(t)$. If the statistic characteristics are known，the statistic characteristics of $a_x(t)$ and $\varphi_x(t)$ or $X_c(t)$ and $X_s(t)$ are also determined.

Suppose $X(t)$ is a stationary narrow band Gaussian random process with 0 mean and δ^2 variance. Now，let us discuss the statistic characteristics of $X_c(t)$ and $X_s(t)$，and then find the statistic characteristics of $a_x(t)$ and $\varphi_x(t)$.

1. Important properties of the in-phase and quadrature components

The schemes in Figure 2.5.3 may be viewed as narrowband noise analyzer and synthesizer.

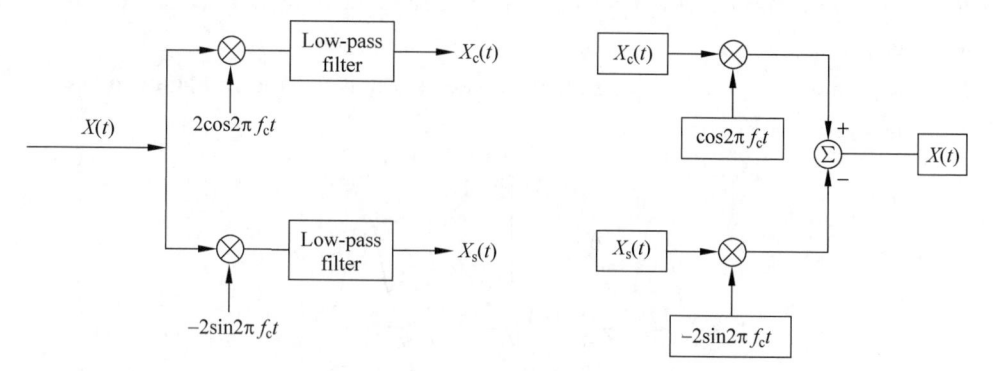

Figure 2.5.3 The analyzer and synthesizer of a narrowband process

Consider a narrowband process $X(t)$ of bandwidth $2B$，each of the two ideal low-pass filter（低通滤波）has a bandwidth equal to B.

The $X_c(t)$ and $X_s(t)$ have some important properties.

(1) The in-phase component $X_c(t)$ and quadrature component $X_s(t)$ have 0 mean and δ^2 variance，that are the same as a narrowband process $X(t)$. $X_c(t)$, $X_s(t)$ are also narrowband Gaussian random processes.

(2) $X_c(t)$ and $X_s(t)$ are uncorrelated and independent at an arbitrary same instant. [Or $X_c(t)$ and $X_s(t)$ at the same instant are uncorrelated and statistically independent.]

(3) Both the in-phase components $X_c(t)$ and quadrature $X_s(t)$ have the same power spectral density (PSD), which is related to the PSD $P_X(f)$ of the narrowband process $X(t)$ as

$$P_{X_c}(f) = P_{X_s}(f) = \begin{cases} P_X(f-f_c) + P_X(f+f_c), & -B \leqslant f \leqslant B \\ 0, & \text{otherwise} \end{cases} \quad (2.5.3)$$

2. Statistic characteristics of $a_x(t)$ and $\varphi_x(t)$

Note that X_c and X_s are independent Gaussian random variables with mean zero and variance δ^2, so we may express their joint probability density function by

$$f(x_c, x_s) = f(x_c)f(x_s) = \frac{1}{2\pi\delta^2} \exp\left(-\frac{x_c^2 + x_s^2}{2\delta^2}\right) \quad (2.5.4)$$

Therefore, the joint probability density function $f(a_x, \varphi_x)$ can be expressed as:

$$f(a_x, \varphi_x) = f(x_c, x_s) \left| \frac{\partial(x_c, x_s)}{\partial(a_x, \varphi_x)} \right| \quad (2.5.5)$$

According to $\begin{cases} x_c = a_x \cos\varphi_x, \\ x_s = a_x \sin\varphi_x \end{cases}$,

$$\left| \frac{\partial(x_c, x_s)}{\partial(a_x, \varphi_x)} \right| = \begin{vmatrix} \dfrac{\partial x_c}{\partial a_x} & \dfrac{\partial x_s}{\partial a_x} \\ \dfrac{\partial x_c}{\partial \varphi_x} & \dfrac{\partial x_s}{\partial \varphi_x} \end{vmatrix} = \begin{vmatrix} \cos\varphi_x & \sin\varphi_x \\ -a_x \sin\varphi_x & a_x \cos\varphi_x \end{vmatrix} = a_x$$

$$f(a_x, \varphi_x) = a_x f(x_c, x_s) = \frac{a_x}{2\pi\delta^2} \exp\left[-\frac{(a_x \cos\varphi_x)^2 + (a_x \sin\varphi_x)^2}{2}\right]$$

$$= \frac{a_x}{2\pi\delta^2} \exp\left(-\frac{a_x^2}{2\delta^2}\right)$$

This probability density function is independent of the phase φ, which means that a_x and φ_x are statistically independent. The mathematics expression is as following:

$$f(a_x, \varphi_x) = f(a_x) \cdot f(\varphi_x)$$

$$f(a_x) = \int_{-\infty}^{\infty} f(a_x, \varphi_x) \mathrm{d}\varphi_x = \frac{a_x}{\delta^2} \exp\left(-\frac{a_x^2}{2\delta^2}\right), \quad a_x \geqslant 0$$

and

$$f(\varphi_x) = \int_{-\infty}^{\infty} f(a_x, \varphi_x) \mathrm{d}a_x = \frac{1}{2\pi}$$

or

$$f(\varphi_x) = \frac{f(a_x, \varphi_x)}{f(a_x)} = \frac{1}{2\pi}, \quad \varphi \in [0, 2\pi]$$

Conclusion:

> The probability density of the envelope of a narrowband process obeys the Rayleigh distribution (瑞利分布), and the probability density of the phase distribution obeys uniform distribution (均匀分布). For one dimensional distribution, $a_x(t)$ and $\varphi_X(t)$ are statistically independent.

2.6 Sine wave plus narrowband Gaussian noise

In many digital and analog modulation systems, the transmission signals and sinusoidal waves are as the modulated signals. When the signals pass through the channel, the noise and interference will be added to the signal.

In order to reduce the noise, a band-pass filter is usually adopted before the demodulator. The output of the band-pass filter is sine wave plus narrowband noise.

The mixed signal can be expressed by

$$r(t) = A\cos(2\pi f_c t) + n(t) \tag{2.6.1}$$

where $n(t) = n_c(t)\cos\omega_c t - n_s(t)\sin\omega_c t$. We assume that $n(t)$ is Gaussian with zero mean and variance σ^2.

A and f_c are both constants, θ is a random phase, in $(0, 2\pi)$, it obeys the uniform distribution

$$
\begin{aligned}
r(t) &= [A\cos\theta + n_c(t)]\cos\omega_c t - [A\sin\theta + n_s(t)]\sin\omega_c t \\
&= Z_c(t)\cos\omega_c t - Z_s(t)\sin\omega_c t \\
&= Z(t)\cos[\omega_c t + \varphi(t)]
\end{aligned}
\tag{2.6.2}
$$

where $Z_c(t) = A\cos\theta + n_c(t)$; $Z_s(t) = A\sin\theta + n_s(t)$.

Here, we neglect the calculation of probability density of $z(t)$ and $\varphi(t)$ and only give the results.

$$f(z) = \frac{z}{\sigma_n^2}\exp\left[-\frac{1}{2\sigma_n^2}(z^2 + A^2)\right] I_0\left(\frac{Az}{\sigma_n^2}\right), \quad z \geqslant 0 \tag{2.6.3}$$

This relation is called the Rician distribution（莱斯分布）. The envelope and phase distributions are shown in Figure 2.6.1 and Figure 2.6.2.

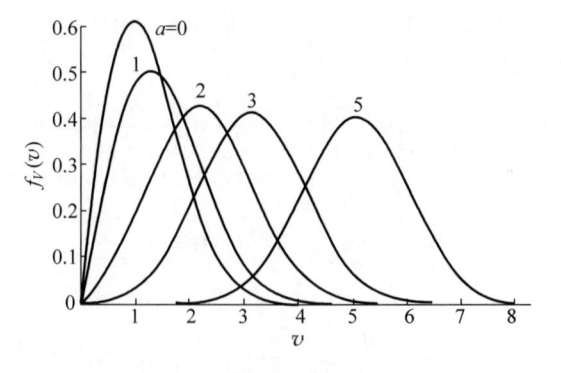

Figure 2.6.1   Normalized Rician distribution

Let $v = \dfrac{z}{\sigma}$, $a = \dfrac{A}{\sigma}$, and $f(v) = \sigma f(z)$, we obtain the normalized form:

$$f(v) = v\exp\left(-\frac{v^2 + a^2}{2}\right) I_0(av) \tag{2.6.4}$$

where $I_0(x)$ is the modified Bessel function of the first kind of zero order（第一类零阶修

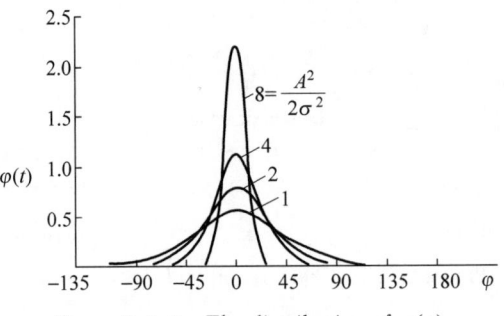

Figure 2.6.2   The distribution of $\varphi(t)$

正贝塞尔函数).

Discussion:

(1) When the signal is week$(A \rightarrow 0)$, then the ratio $\nu = \dfrac{A^2}{2\sigma^2} \rightarrow 0$, $\rightarrow I_0(x) = 1$, the Rician distribution reduces to the Rayleigh distribution. The phase distribution is uniform distribution.

(2) When the signal to noise ratio $\nu = \dfrac{A^2}{2\sigma^2}$ is strong/big, the envelope distribution is approximately Gaussian.

2.7   Gaussian white noise and band pass white noise

视频讲解

Noise can be classified into impulse noise, narrow band noise and fluctuation noise. When the influence of noise on communication systems is discussed, the influence of the fluctuation noise, especially thermal noise should be mainly considered. Thermal noise is usually called the white noise, because its frequency spectrum is uniformly distributed, which is similar to the spectrum of white light in the frequency range of the visible light.

1. White noise

If the PSD of noise is a constant, then the noise is called white noise $n(t)$.

$$P(f) = \frac{n_0}{2}, \quad -\infty < f < +\infty (\text{W/Hz}) \quad \text{(Double sideband)}$$

$$P(f) = n_0, \quad 0 < f < +\infty (\text{W/Hz}) \qquad \text{(Single sideband)}$$

where n_0 is a constant. The PSD and auto-correlation distributions are shown in Figure 2.7.1. They are Fourier transform pair.

2. Low-pass white noise

The sketch of white noise passing the low-pass filter is shown in Figure 2.7.2.

If the input is white noise, the output of low-pass filter is called low-pass noise, and its corresponding PSD is

$$P_{n_0}(f) = \begin{cases} \dfrac{n_0}{2}, & |f| \leqslant B \\ 0, & \text{others} \end{cases}$$

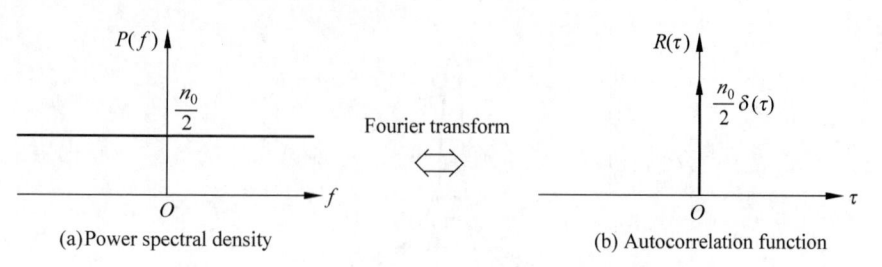

(a) Power spectral density   (b) Autocorrelation function

Figure 2. 7. 1　Characteristics of the white noise

Figure 2. 7. 2　Low-pass filter

and the autocorrelation function is：

$$R(\tau) = n_0 B \frac{\sin 2\pi B\tau}{2\pi B\tau}$$

Their curves are drew in Figure 2. 7. 3.

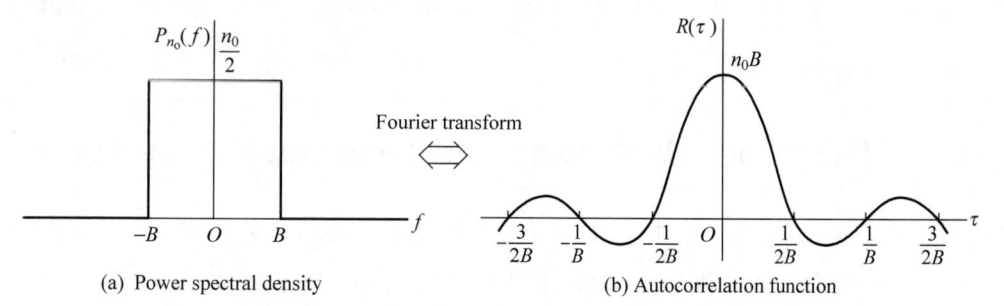

(a) Power spectral density   (b) Autocorrelation function

Figure 2. 7. 3　Characteristics of low-pass filtered white noise

3. Band-pass white noise

The sketch of white noise passing the band-pass filter is shown in Figure 2. 7. 4.

Figure 2. 7. 4　Band-pass filter

The transmission characteristic of $H(f)$ is

$$H(f) = \begin{cases} 1, & f_c - \dfrac{B}{2} \leqslant |f| \leqslant f_c + \dfrac{B}{2} \\ 0, & \text{otherwise} \end{cases}$$

where f_c is the carrier wave frequency，B is the bandwidth.

The PSD of white noise after passing through the BPF is

$$P_{n_o}(f) = \begin{cases} \dfrac{n_0}{2}, & f_c - \dfrac{B}{2} \leqslant |f| \leqslant f_c + \dfrac{B}{2} \\ 0, & \text{otherwise} \end{cases}$$

and the autocorrelation function is

$$R(\tau) = \int_{-\infty}^{\infty} P_{n_o}(f) e^{j2\pi f_c} \mathrm{d}f = n_0 B \frac{\sin\pi B\tau}{\pi B\tau} \cos 2\pi f_c \tau$$

when $\quad B \ll f_c \quad$ and $\quad f_c \gg 0$

This bandwidth of the white noise is also called the narrowband Gaussian white noise. Its corresponding PSD and autocorrelation curves are in Figure 2.7.5.

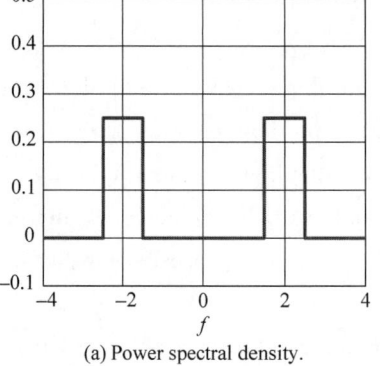
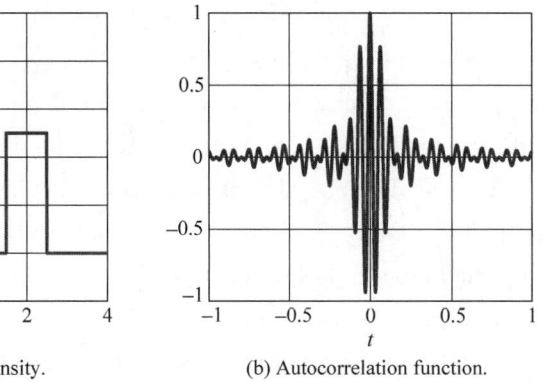

(a) Power spectral density.　　　(b) Autocorrelation function.

Figure 2.7.5　Characteristics of the band-pass filtered white noise

Summary and discussion

Both signal and noise in communication systems can be regarded as random processes varying with time, especially the stationary and ergodic processes.

The random processes have the characteristics of the random variables and functions of time, which can be described from two different and correlative aspects:

(1) The sample space or the ensemble composed of sample functions;

(2) An indexed ensemble (family) of random variables.

The characteristics of random signals can be expressed by probability distribution and probability density functions. If the statistic characteristic of a random process is independent of the time origin, then the random process is called a strict-sense stationary (严平稳) random process; Another important concise method for the partial description of random signals is the numerical characteristics. If the statistic characteristics of a random process is independent of the time origin, then the random process is called a wide-sense stationary (广义平稳) random process. Two ensemble-average parameters should satisfy the following conditions:

(1) Mean is independent of time;

(2) Autocorrelation function depends only on the time difference.

If the statistical mean of a random process is equal to its time average, then the random process has ergodicity. The statistical mean may be replaced by the time average of an arbitrary realization of the ergodic process.

Another important parameter of a random process is the power spectral density (PSD). The autocorrelation function and the PSD constitute a Fourier-transform pair. Autocorrelation function and PSD are two important numerical characteristics to describe the stationary random processes.

If the input to an LTI filter is a stationary process, then the output is also a stationary process. The relationship is as following:

$$E[Y(t)] = \mu_Y \cdot H(0) = \text{constant}$$
$$P_Y(f) = |H(f)|^2 P_X(f)$$

The thermal noise in communication systems is generally a Gaussian process, which is also called the normal random process. Its numerical characteristics can completely describe this process. The one dimensional probability density function only is determined by the mean and variance. The two dimensional probability density function is mainly determined by the autocorrelation function. The Gaussian process after passing through a linear system remains a Gaussian process.

The statistical characteristics of narrowband random process are usually used to analyze the modulated system, pass band system or wireless multi-path channel. According to the definition and two representations of the narrowband signal, the envelope and phase distribution of narrow band noise and sinusoidal wave plus narrow band Gaussian process can be derived.

There are three kinds of stationary random processes which often exist in communication systems:

(1) Gaussian process represented by the thermal noise, which is also called white noise. When Gaussian white noise is passed through low-pass or band-pass filter, the low-pass Gaussian white noise or band-pass Gaussian white noise is generated, respectively. Their autocorrelation and PSD should be remembered, which will be used in the following chapter.

(2) Rayleigh distribution process represented by the envelope of narrow band noise.

(3) Rician distribution process represented by the envelope of sinusoidal wave plus narrow band Gaussian process.

Homework

2.1 A random telegraph signal $X(t)$, characterized by the autocorrelation function
$$R_X(t) = \exp(-2v|\tau|)$$
where v is a constant, is applied to the low-pass RC filter of the following Figure 2.1. Determine the PSD of the random process at the filter output.

2.2 Let X_1 and X_2 be statistically independent Gaussian-distributed random variables, each with zero mean and variance δ^2. Define the Gaussian process

$$Y(t) = X_1 \cos(\omega_0 t) - X_2 \sin(\omega_0 t)$$

(1) Find $E[Y(t)], E[Y^2(t)]$.

(2) The probability density function of $Y(t)$.

(3) $R(t_1, t_2)$.

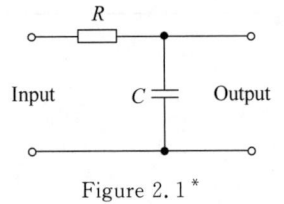

Figure 2. 1*

2.3　If a Gaussian white noise passes through the filter shown in Fig. 2. 1. Its mean is 0 and the double-side PSD is $n_0/2$. Find the PSD and autocorrelation of the output noise.

Vocabulary

algebraic	代数	Fourier transform	傅里叶变换
arbitrary	任意的	integral	积分
cofactor	余子式	interference	干扰
complementary	补充的	matrix	矩阵
correlation	相关	multipath	多路径的
covariance	协方差	nonnegative	非负
convolution	卷积	phase	相位
density	密度	quadrature	正交
deviation	偏差	random	随机的
dimension	维度	sinusoidal	正弦的
deterministic	确定性的	stationary	平稳的
equation	方程	statistical	统计学的
envelope	包络	stochastic	随机的
ergodicity	各态历经性	variance	方差
frequency spectrum	频谱	uniformly	均匀地

---

* 书中电路符号采用了英文图书的常见形式,未按国标修改。

Channel

Mind map:

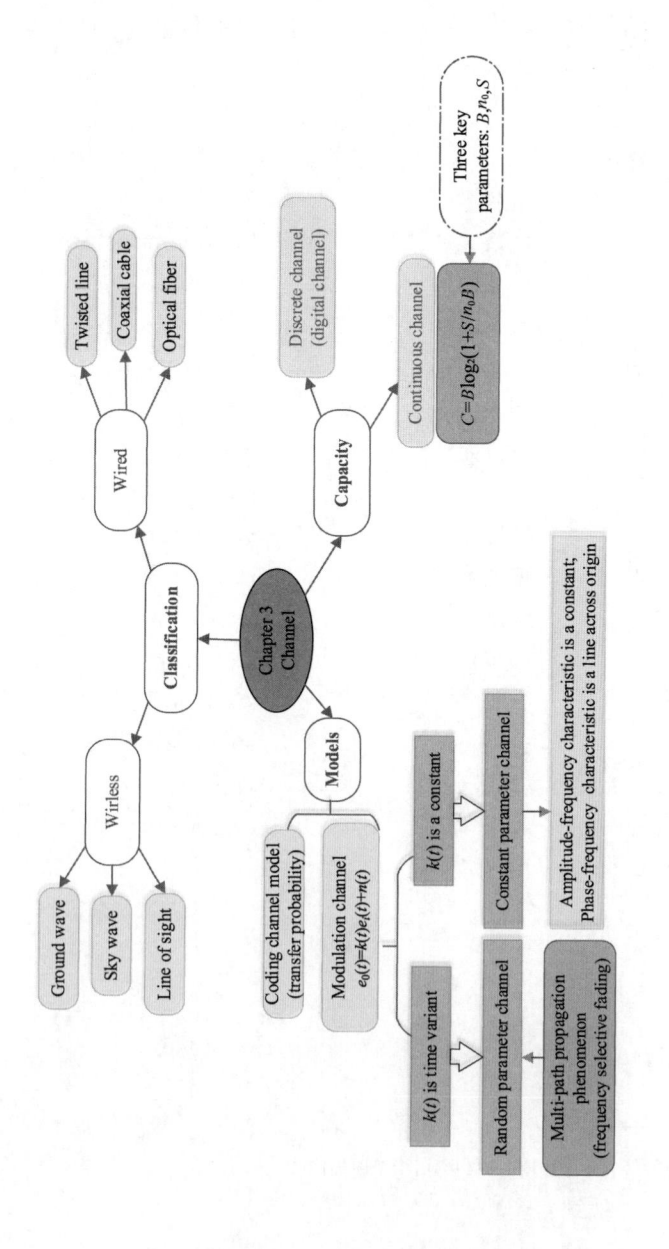

Mind map for
Chapter 3

In this chapter, the classification of the channel, the channel models, the noise in the channel and the capacity of channel will be introduced and discussed.

3.1　The classification of channels

According to the transmitted media, channels can be classified into two categories: wireless channel and wired channel.

3.1.1　Wireless channels

The transmission of signals in wireless channel is achieved by the propagation of electromagnetic waves in space. In principles, the electromagnetic wave of any frequency can be produced. For effectively transmitting and receiving, the frequency of the magnetic wave used for communications is usually rather high in practical applications.

According to different communication ranges, frequencies and locations, the electromagnetic wave propagation can be classified into three types: the line of sight (LOS: 视线) propagation, the ground wave(地波) and the sky wave(天波)(or called ionosphere reflection wave: 电离层反射波),as shown in Table 3.1.1. The sketch maps of these three propagations are shown in Figure 3.1.1.

Table 3.1.1　The comparison of three propagation modes

Type	Frequency range	Propagation mode	Propagation range
Ground wave	< 2MHz	Along the curved ground surface	Over hundreds to thousands of km
Sky wave	$2\sim30$MHz	Many reflections between ground and flayer	More than 10 000km
LOS	>30MHz	Line of sight	$D=\sqrt{50h}$ (km) $\rightarrow$Satellite communication

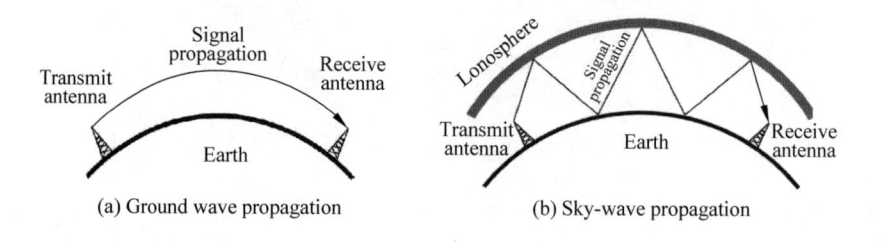

(a) Ground wave propagation　　　(b) Sky-wave propagation

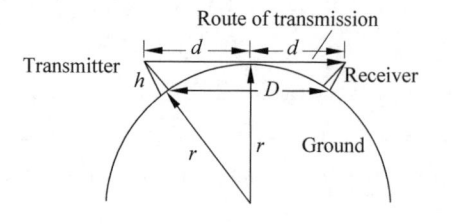

(c) LOS propagation

Figure 3.1.1　Sketch of three propagations

The propagation of electromagnetic wave in the atmosphere is influenced by the atmosphere. The relationship between the attenuation characteristics of the atmosphere and the frequency are shown in the following Figure 3.1.2.

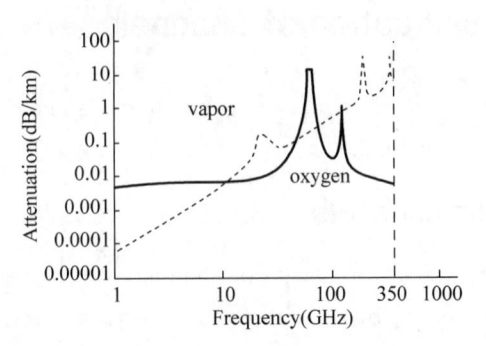

Figure 3.1.2   Attenuation of oxygen(氧气) and vapor(水蒸气)(concentration 7.5g/m³)

3.1.2 Wire channel

There are three kinds of wired channels: symmetrical cables, coaxial cables and optical fibers.

Symmetrical cables is also called twist wire（双绞线）. The telephone channel is built using twist pairs for signal transmission. A twisted pair consists of two solid copper conductors, each of which is encased in a polyvinylchloride（PVC：聚氯乙烯）sheath. Twisted pairs are usually made up into cables, as in Figure 3.1.3, with each cable consisting of many pairs in close proximity to each other. Twisted pairs are naturally susceptible to electromagnetic interference（EMI：电磁干扰）.

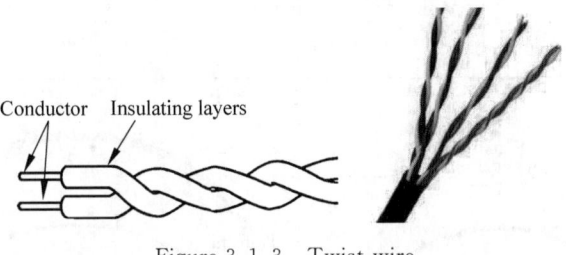

Figure 3.1.3   Twist wire

Typical coaxial cable（同轴电缆）has a characteristic impedance of 50 or 75 ohms. The composition of coaxial cable is shown in Figure 3.1.4. Compared to a twisted-pair cable, a coaxial cable offers a greater degree of immunity to EMI. The standard bite rate is 10Mb/s, which is higher than twisted pairs. The applications of coaxial cables are as the transmission medium for local area networks（LAN）and in cable-television systems.

The optical fiber is widely used for the transmission of light signals from one place to another. It can be classified into two categories: multi-mode and single-mode, as given in Figure 3.1.5.

Optical fibers have unique characteristic that make them highly attractive as a

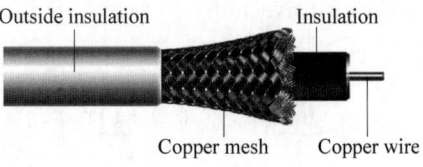

Figure 3.1.4 Coaxial cable

transmission medium. They offer the following unique characteristics:

- Enormous potential bandwidth.
- Low transmission losses.
- Immunity to EMI.
- Small size and weight.
- Ruggedness and flexibility.

Figure 3.1.5 The structure of single-mode and multimode optical fiber

There are two minimum loss points at $1.31\,\mu m$ and $1.55\,\mu m$ from the following Figure 3.1.6. Therefore, these two wavelengths are widely used.

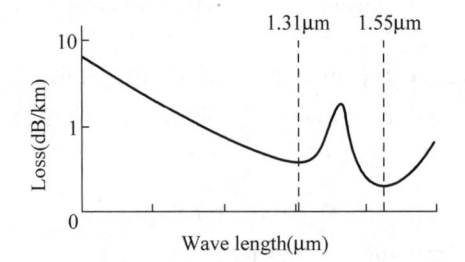

Figure 3.1.6 The relationship between loss and wavelength

3.2 Channel models

视频讲解

We have introduced two types of channels, how to describe the channels in mathematical tool is the content of this part. There are two basic channel models: one is for modulation, another one is for coding.

3.2.1 Modulation channel model（调制解调模型）

The basic modulation channel is defined as

$$e_o(t) = f[e_i(t)] + n(t) \tag{3.2.1}$$

where $e_i(t)$ is the signal voltage at the channel input terminal, $e_o(t)$ is the signal voltage at the channel output, and $n(t)$ is the noise voltage. Noise $n(t)$ always exists in the channel. It is usually called the additive noise because "+". The model is illustrated in Figure 3.2.1.

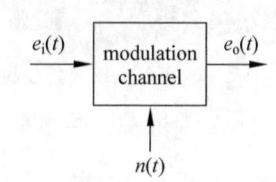

Figure 3.2.1   Modulation channel model

$f(\cdot)$ is the function between $e_i(t)$ and $e_o(t)$. For simplicity, we usually assume $f[e_i(t)]=k(t)e_i(t)$.

$k(t)$ is a complicated function and it reflects the characteristics of the channel.

$$k(t)=\begin{cases} \text{time variant} \Rightarrow \text{random parameter channels(随参信道)} \\ \text{constant} \Rightarrow \text{constant parameter channels(恒参信道)} \end{cases}$$

3.2.2   Coding channel model（编码信道模型）

The input and output signals of the coding channel are digital sequences in Figure 3.2.2.

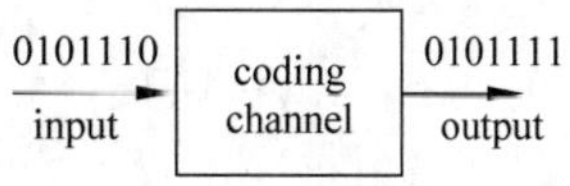

Figure 3.2.2   Coding channel

Error usually happens at the output because of interference. Therefore, the best method to describe this model is the error probability（错误概率）. It is also called the transfer probability.

$$\left.\begin{array}{l} P(0/0) \quad\quad \text{transmitting } 0 \quad \text{and } \text{receiving } 0 \\ P(1/1) \quad\quad \text{transmitting } 1 \quad \text{and } \text{receiving } 1 \end{array}\right\} \text{Correct transfer probability}$$

$$\left.\begin{array}{l} P(1/0) \quad\quad \text{transmitting } 0 \quad \text{and } \text{receiving } 1 \\ P(0/1) \quad\quad \text{transmitting } 1 \quad \text{and } \text{receiving } 0 \end{array}\right\} \text{Error transfer probability}$$

For binary systems：

$$P(0/0)=1-P(1/0)$$
$$P(1/1)=1-P(0/1)$$

The model in Figure 3.2.3 is the simple binary coding memoryless channel model, in which the occurrence of errors in adjacent symbols is independent.

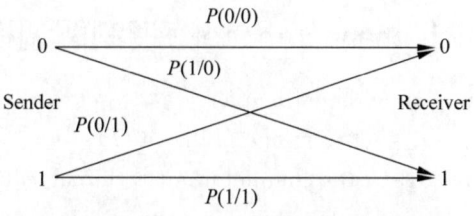

Figure 3.2.3   Binary coding channel model

3.3   Influence of the channel characteristics on transmission（for modulation model）

3.3.1   Influence of constant parameter channel on signal transmission

The main transmission characteristics of the transmission function are usually described by the amplitude-frequency characteristics（幅频特性）and phase-frequency characteristics （相频特性）.

In practice，phase-frequency characteristic can also be described by group delay（群延迟）.

The amplitude characteristic can be described by insertion loss（插入损耗）. Figure 3.3.1 show the ideal amplitude-frequency and phase-frequency characteristics.

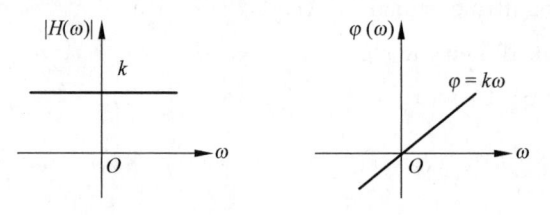

Figure 3.3.1   The ideal amplitude-frequency and phase-frequency characteristics

The definition of group delay is

$$\tau(\omega) = \frac{\mathrm{d}\varphi(\omega)}{\mathrm{d}\omega}$$

The plots of Figure 3.3.2 clearly illustrate the dispersive nature of the telephone channel.

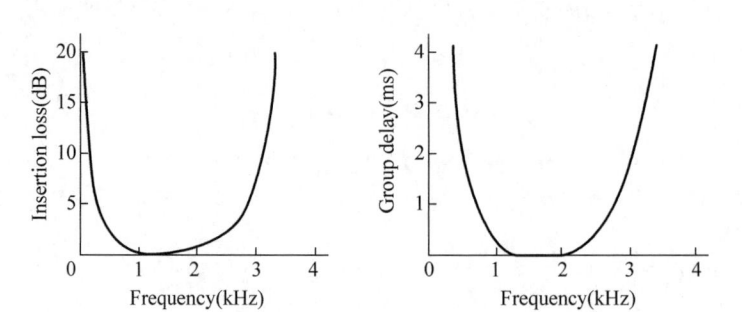

Figure 3.3.2   Characteristic of typical telephone connection

3.3.2   Influence of random parameter channel of signal transmission

There are there common characteristics：

(1) Transmission attenuation of the signal is varying with time；

(2) Transmission delay of the signal varies with time；

(3) Signal arrives at the receiver over several paths，i. e. multi-path propagation

phenomenon exists.

Multi-path will be discussed for its great influence on the quality of the signal transmission, the model of multipath is illustrated in Figure 3.3.3.

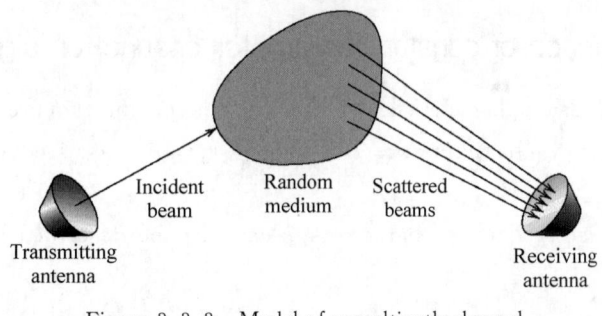

Figure 3.3.3　Model of a multipath channel

Suppose the transmitting signal is $A\cos\omega_0 t$, which is a common signal model in a communication system. When the signal propagates to the receiver over n paths, then the received signal $R(t)$ may be written as:

$$R(t) = \sum_{i=1}^{n} \mu_i(t)\cos\omega_0[t - \tau_i(t)] = \sum_{i=1}^{n} \mu_i(t)\cos[\omega_0 t + \varphi_i(t)] \qquad (3.3.1)$$

where

$$\left.\begin{array}{l} \mu_i(t) \text{ is the attenuation of } i\text{-th path} \\ \tau_i(t) \text{ is the delay of } i\text{-th path} \\ \varphi_i(t) = -\omega_0 \tau_i(t) \end{array}\right\} \text{ are random varying}$$

Equation (3.3.1) can also be written as:

$$R(t) = \sum_{i=1}^{n} \mu_i(t)\cos\varphi_i(t)\cos\omega_0 t - \sum_{i=1}^{n} \mu_i(t)\sin\varphi_i(t)\sin\omega_0 t \qquad (3.3.2)$$

Let

$$X_c(t) = \sum \mu_i \cos\varphi_i(t)$$

$$X_s(t) = \sum \mu_i \sin\varphi_i(t)$$

Then

$$R(t) = X_c(t)\cos\omega_0 t - X_s(t)\sin\omega_0 t = V(t)\cos[\omega_0 t + \varphi(t)]$$

where

$$V(t) = \sqrt{X_c^2(t) + X_s^2(t)} \text{ —envelope}$$

$$\varphi(t) = \arctan\frac{X_s(t)}{X_c(t)} \text{ —phase}$$

Comparing this eq. with the narrowband random process yields:

$$X(t) = X_c\cos\omega_c t - X_s\sin\omega_c t = a_x(t)\cos[\omega_c t + \varphi_x(t)]$$

So $R(t)$ can be regarded as a narrowband signal with random varying envelope and phase, as shown in Figure 3.3.4.

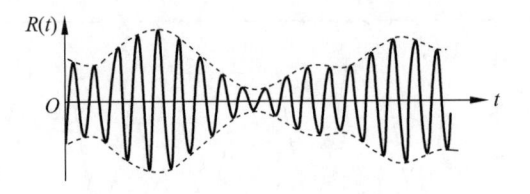

Figure 3.3.4　The narrowband signal

To simplify the problem, here we only discuss two paths of the multi-path propagation with the same attenuation and different delays.

Suppose the transmission signal is $f(t)$, the received signals are $Af(t-\tau_0)$ and $Af(t-\tau_0-\tau)$, respectively, where A is a constant.

Their corresponding Fourier transforms are

Input:

$$f(t) \Leftrightarrow F(\omega)$$

$$Af(t-\tau_0) \Leftrightarrow AF(\omega)e^{-j\omega\tau_0}$$

$$Af(t-\tau_0-\tau) \Leftrightarrow AF(\omega)e^{-j\omega(\tau_0+\tau)}$$

Output:

$$Af(t-\tau_0) + Af(t-\tau_0-\tau) \Leftrightarrow AF(\omega)e^{-j\omega\tau_0}(1+e^{-j\omega\tau})$$

Therefore, the transfer function of the two paths channel is

$$H(\omega) = Ae^{-j\omega\tau_0}(1+e^{-j\omega\tau})$$

$$|H(\omega)| = |Ae^{-j\omega\tau_0}(1+e^{-j\omega\tau})|$$

$$= A|1+e^{-j\omega\tau}|$$

$$= A|1+\cos\omega\tau - j\sin\omega\tau| = A\sqrt{(1+\cos\omega\tau)^2 + \sin^2\omega\tau} = 2A\left|\cos\frac{\omega\tau}{2}\right|$$

The curve drawn according to the above equation is shown in Figure 3.3.5.

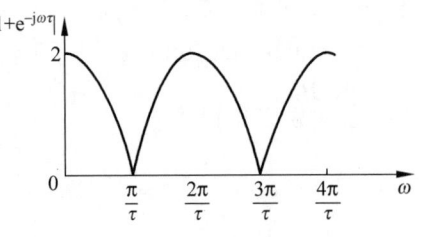

Figure 3.3.5　Multipath effect

From the above curve, we consider that fading is related to the frequency, it is called **frequency selective fading**（频率选择性衰落）.

Figure 3.3.6 illustrates the effect of Rayleigh fading on the waveform of the received signal, whose amplitude and phase components vary randomly with time.

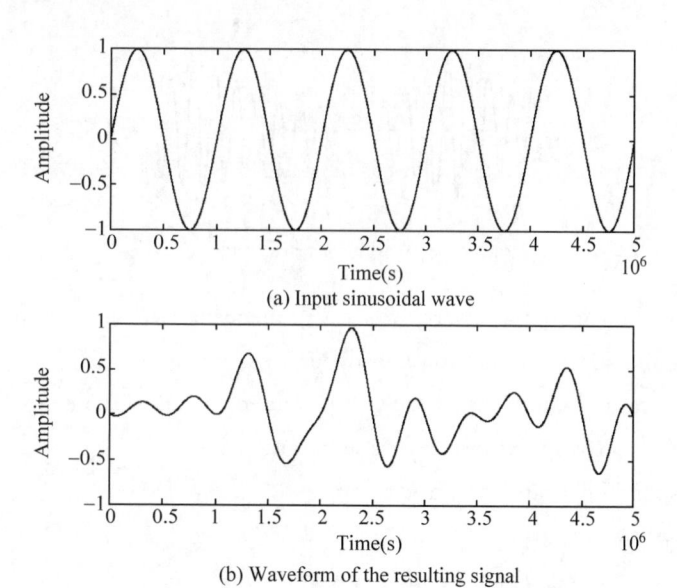

(a) Input sinusoidal wave

(b) Waveform of the resulting signal

Figure 3.3.6　Effect of Rayleigh fading on a sinusoidal wave

视频讲解

3.4　Channel capacity（continuous channel）

The information capacity of a continuous channel of bandwidth $B\,\mathrm{Hz}$, with the addictive white Gaussian noise of PSD $\dfrac{n_0}{2}$ and limited in B, is given by

$$C_t = B\log_2\left(1 + \frac{S}{n_0 B}\right)\,(\mathrm{b/s}) \tag{3.4.1}$$

where S is the average transmitted power.

There are three key system parameters: channel bandwidth (B), average transmitted power (S) and the noise power spectral density $(n_0/2)$.

(1) When S increases or n_0 decreases, C increases.

(2) When $B \to \infty$, C_t approaches to the following limit:

$$\lim_{B \to \infty} C_t = \lim_{x \to 0} \frac{S}{n_0}\frac{Bn_0}{S}\log_2\left(1 + \frac{S}{n_0 B}\right)$$

$$= \lim_{x \to 0} \frac{S}{n_0}\log_2(1 + x)^{1/x} = \frac{S}{n_0}\log_2 e \approx 1.44\frac{S}{n_0}$$

The channel capacity approaches 1.44 times of signal power to noise PSD ratio.

Example 3.4.1: A frame of black and white TV image is composed of 300 thousand pixels, each pixel has 10 levels of brightness and these 10 levels occur at equal probabilities. If the image is transmitted at a rate of 25 frames per second, the image signal to noise ratio is required to reach 30dB, find the required transmission bandwidth.

Answer: Since each pixel takes 10 possible levels with equal probability, the information content of each pixel I_p is

$$I_p = \log_2 10 = 3.32 (\text{b/pixel})$$

The information content I_f of each image frame is

$$I_f = 30\,000 \times 3.32 = 9.96 \times 10^5 (\text{b/frame})$$

Since there are 25 frames of image per second, the required information transmission rate is

$$I_f \times 25 = 9.96 \times 25 \times 10^5 = 24.9 \times 10^6 (\text{b/s})$$

According to the formula

$$C = B \log_2 (1 + S/N)$$

$$24.9 \times 10^6 = B \log_2 (1 + 10^3) = 9.96B$$

The bandwidth is:

$$B = 24.9 \times 10^6 / 9.96 = 2.5 (\text{MHz})$$

Summary and discussion

Channel plays a very important role in the communication system, which also is the basis of the following chapters. The classification of the channel is given in this chapter. Three main wireless channels are the line of sight (LOS: 视线) propagation, the ground wave(地波) and the sky-wave(天波). There are also three main kinds of wired channels: symmetrical cables, coaxial cables and optical fibers. Nowadays, these channels are widely used, and they have affected our daily lives, due to their wide range of applications.

It is important to understand two mathematics models of channel. Modulation channel model is usually regarded as analog channel. The multiplicative noise and additive noise are used to reflect the channel effect. Multiplicative noise $k(t)$ can cause signal distortion, including linear distortion, non-linear distortion, time delay and attenuation. Depending on the $k(t)$ being a constant or time varying, constant parameter channel and random parameter channel are distinguished. The influence of random parameter channel on signal is multipath effect, which can cause the frequency selective fading of the signal. Additive noise always exists in communication systems. Thermal noise is usually called white noise. We mainly discuss the influence of the white noise, especially the Gaussian white noise in this book.

Coding channel model is usually viewed as digital channel. Both additive noise and multiplicative noise influence the coding channel. Error probability is used to describe this kind of channels, which is also called the transfer probability.

There is a Shannon-Hartley theorem (香农定理) about continuous channel capacity:

$$C_t = B \log_2 \left(1 + \frac{S}{n_0 B}\right) (\text{b/s})$$

C_t represents maximal information rate which can be transmitted by the channel, and its unit is b/s. The bandwidth and SNR can be traded off. If the bandwidth is increased then the SNR decreases and the capacity remains unchanged. This trade-off relationship has very steering significance in the design of communication systems. This trade off cannot be naturally achieved. It required that the signal is modulated or encoded to increase its occupied bandwidth，and then it is sent to the channel for transmission corresponding to demodulation for decoding at the receiver.

Homework

3. 1　Assume a wireless link uses line-of-sight propagation for communication，and the heights of the transmitting antenna and the receiving antenna are both 80m. Find the maximum communication distance.

3. 2　A voice-grade channel of the telephone network has a bandwidth of 3. 4kHz.

（1）Calculate the information capacity of the telephone channel for a signal-to-noise ratio of 30dB.

（2） Calculate the minimum SNR required to support information transmission through the telephone channel at the rate of 9600b/s.

Vocabulary and terminologies

attenuation	衰减	coaxial cable	同轴电缆
copper	铜	constant parameter channel	恒参信道
channel	信道	EMI	电磁干扰
electromagnetic	电磁的	frequency selective fading	频率选择性衰落
fiber	光纤	ground wave	地波
interference	干扰	group delay	群延时
oxygen	氧气	insert loss	插入损耗
propagation	传播	multipath effect	多径效应
surface	表面	random parameter channel	随参信道
terminal	终端	sky wave	天波
vapor	水蒸气	transfer probability	转移概率
wireless	无线	twist wire	双绞线
channel capacity	信道容量		

Chapter 4 Continuous-wave modulation（analog modulation system）

In this chapter，we study continuous-wave modulation，which is basic to the operation of analog communication systems. There are two kinds of modulation：linear modulation （线性调制）and nonlinear modulation.

- Linear modulation：amplitude modulation（AM）

$$\begin{cases} \text{double-sideband（suppressed carrier）（DSB-SC）modulation} \\ \text{single-sideband（SSB）modulation} \\ \text{vestigital-sideband（VSB）modulation} \end{cases}$$

- Nonlinear modulation：

$$\begin{cases} \text{frequency modulation（FM）} \\ \text{phase modulation（PM）} \end{cases}$$

The anti-noise performances of different modulations are also compared and discussed.

4.1 Introduction

视频讲解

In the continuous wave modulation system，the input singal is analog. As depicted in Figure 4.1.1，we can find that the transmitter consists of a modulator and the receiver consists of a demodulator.

Information source → Modulator → Channel → Demodulator → Information destination

Figure 4.1.1 Continuous-wave modulation system

Modulation：the process by which some characteristics of a carrier is varied in accordance with a modulating wave signal.

Demodulation：the reverse of modulation process.

Message signal：the information-bearing signals are also referred to as baseband signals.

Carrier wave：is cosine waveform in this chapter.

$$c(t) = A\cos(\omega_c t + \varphi)$$

A carrier has three parameters: amplitude *A*, angular frequency ω_c and initial phase φ.

> AM: the **amplitude** of a cosine carrier is varied in accordance with $m(t)$.
> FM: the **frequency** of a cosine carrier is varied in accordance with $m(t)$.
> PM: the **phase** of a cosine carrier is varied in accordance with $m(t)$.

Modulated signal $s_m(t)$（已调信号）: the carrier after being modulated is called modulated signal.

Modulator（调制器）: the device for modulation is called the modulator.

The function and purpose of modulation

1. Frequency spectrum shift

The proper use of the communication channel requires a shift of the range of baseband frequencies into other frequency ranges suitable for transmission, and a corresponding shift back to the original frequency range after reception.

2. Anti-noise performance

The anti-noise performance can be improved by modulation.

Although the digital communication is superior to the analog communication and develops rapidly, the analog modulation is still the basic modulation mode. Some concepts and principles of analog modulation are also used in digital communication.

4.2 Linear modulation

The basic model of a linear modulator is shown in Figure 4.2.1.

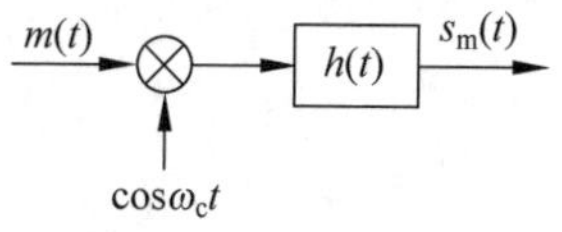

The modulating signal $m(t)$ and the carrier are multiplied, then it passes through a bandpass filter $h(t)$.

Figure 4.2.1 Basic model of a linear modulator

The characteristics of $H(f)$ may have different designs, so different modulation modes are resulted. They will be introduced in the following part. Take the SSB modulation as an example, as illustrated in Figure 4.2.2, the spectrum after modulation is the frequency shift of the message signal $m(t)$.

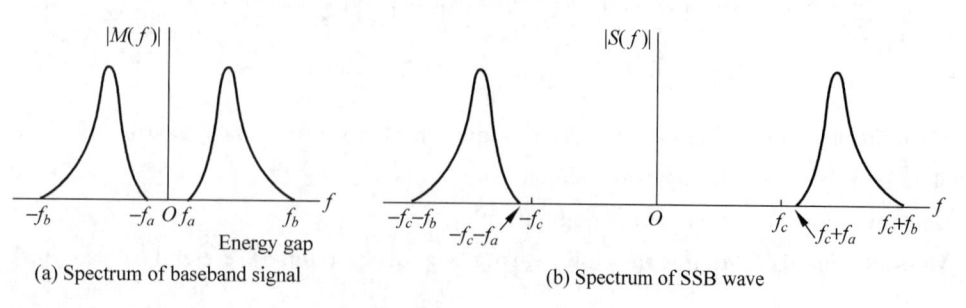

(a) Spectrum of baseband signal

(b) Spectrum of SSB wave

Figure 4.2.2 The spectrum shifting of SSB

4.2.1   AM

Assume $m(t)$ contains D. C. component (直流分量).

If $H(f)=C$, it is an all-pass filter(全通滤波器), then this linear modulation is amplitude modulation, as shown in the Figure 4.2.3.

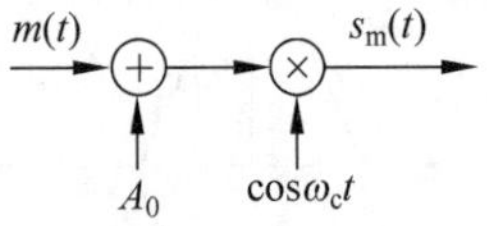

$$s_{AM}(t)=[A_0+m(t)]\cos\omega_c t$$
$$=A_0\cos\omega_c t+m(t)\cos\omega_c t \qquad (4.2.1)$$

In Figure 4.2.4, when $|m(t)|_{max}>A_0$ for any t, the carrier wave becomes over-modulated, resulting in carrier phase reversals(反相) at the crossed zero points(过零点). The modulated wave then exhibits envelope distortion.

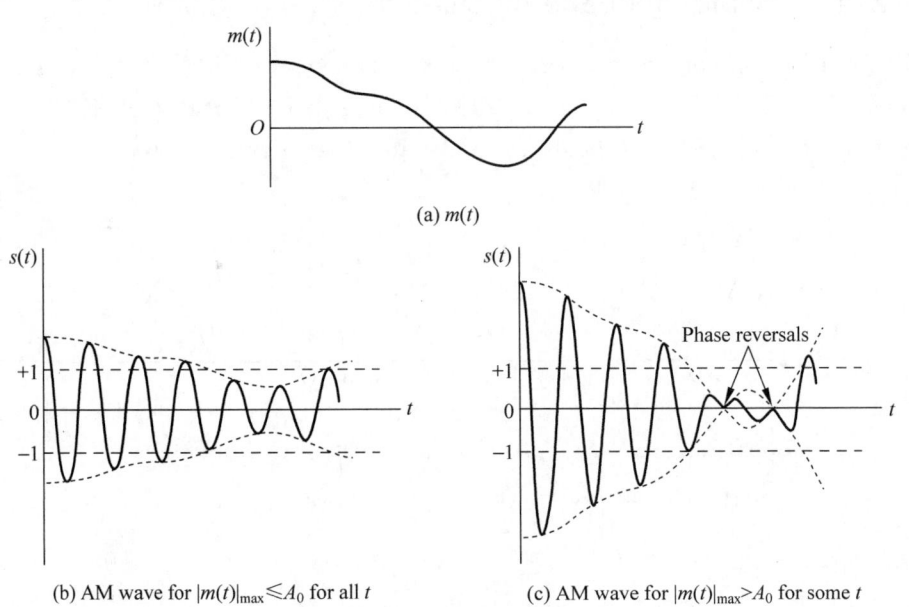

Figure 4.2.3   The model of
AM modulator

(a) $m(t)$

(b) AM wave for $|m(t)|_{max}{\leqslant}A_0$ for all t (c) AM wave for $|m(t)|_{max}{>}A_0$ for some t

Figure 4.2.4   AM modulation

From equation (4.2.1), we can find the Fourier transform of the AM wave $s_m(t)$:

$$S_{AM}(\omega)=\pi A_0[\delta(\omega+\omega_c)+\delta(\omega-\omega_c)]+\frac{1}{2}[M(\omega+\omega_c)+M(\omega-\omega_c)] \qquad (4.2.2)$$

It is not difficult to know from the waveform of AM signal, that shape of the envelope of the modulated signal is the same as the shape of the modulated signal. Therefore, envelope detection method can be used to restore the original modulated signal during demodulation at the receiver. As can be seen from Figure 4.2.5, most part of the power in such a modulated signal is occupied by the carrier, and the carrier itself does not contain the information of the baseband signal. Therefore, the carrier is not necessary for transmission. Another limitation is wasteful of bandwidth. The upper and lower sidebands are symmetry about carrier f_c. This means only one sideband is necessary.

To overcome these limitations, we must make certain modifications: suppress the

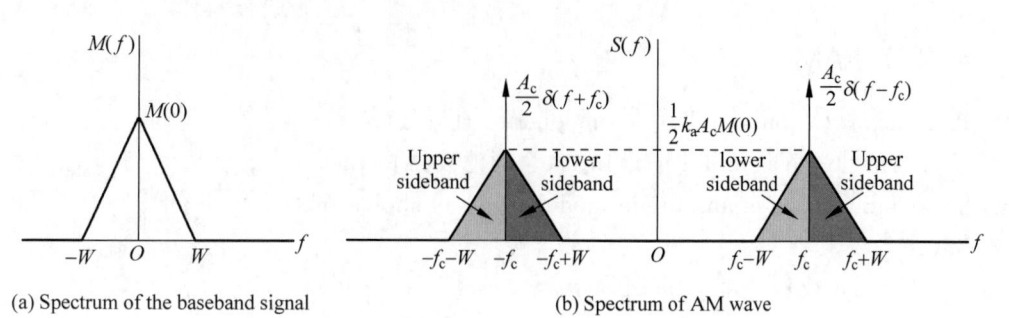

(a) Spectrum of the baseband signal   (b) Spectrum of AM wave

Figure 4.2.5　The spectrum shifting of AM

carrier and modify the sidebands of AM wave, which will be discussed in next section.

4.2.2　Double-sideband modulation

If there is no DC component, then there will be no carrier in the output signal of the multiplier. This kind of modulation is called double-sideband(DSB,双边带)modulation, where the upper and lower sidebands are transmitted(see Figure 4.2.6).

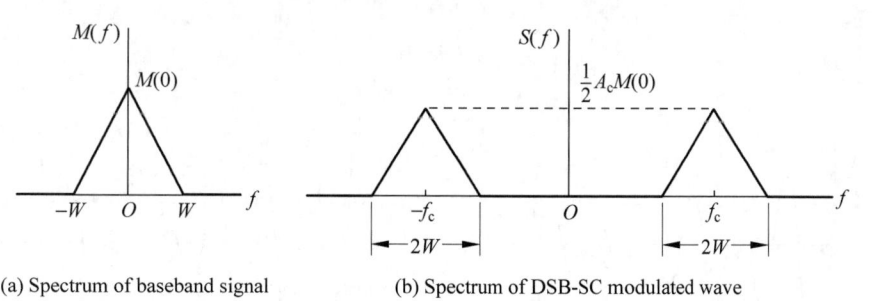

(a) Spectrum of baseband signal   (b) Spectrum of DSB-SC modulated wave

Figure 4.2.6　The spectrum shifting of DSB

We write:

$$s_{\mathrm{DSB}}(t) = m(t)\cos\omega_c t \qquad (4.2.3)$$

and its Fourier transform is obtained as:

$$S(f) = \frac{1}{2}[M(f - f_c) + M(f + f_c)] \qquad (4.2.4)$$

The modulated signal $s_{\mathrm{DSB}}(t)$ has phase reversal when the message signal $m(t)$ crosses zero.

The envelope of DSB-SC modulated signal is different from the message signal. In order to recovery $m(t)$, the coherent detection is usually used at the receiver, as shown in Figure 4.2.7.

It consists of a multiplier and a low-pass filter. The local oscillator is exactly coherent of synchronized with the carrier wave $C(t)$. This method of demodulation is known as coherent detection（相干检测）or synchronous demodulation（同步解调）.

$$s_{\mathrm{m}}(t)\cos\omega_c t = m(t)\cos^2\omega_c t = \frac{1}{2}m(t) + \frac{1}{2}m(t)\cos 2\omega_c t \qquad (4.2.5)$$

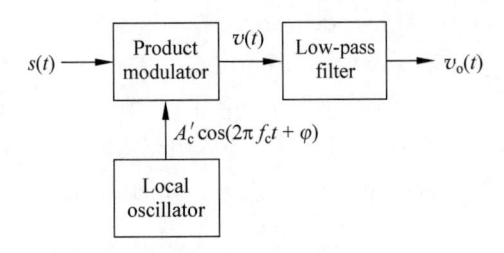

Figure 4. 2. 7   Coherent detection of DSB

After the LPF, the output is

$$m_{\circ}(t) = \frac{1}{2} m(t) \qquad (4.2.6)$$

4.2.3   Single-sideband modulation

In single-sideband(SSB, 单边带) modulation, only the upper or lower sideband is transmitted. A band-pass filter (BPF) needs to be designed to pass one of the sidebands of DSB modulated wave and suppress the other.

A high-pass filter with transfer function $H_{USB}(\omega)$ used in the modulator can pass the upper-sideband signal.

$$H(\omega) = H_{USB}(\omega) = \begin{cases} 1, & |\omega| > \omega_c \\ 0, & |\omega| \leqslant \omega_c \end{cases} \qquad (4.2.7)$$

A low-pass filter with transfer function $H_{LSB}(\omega)$ can pass the lower-sideband signal.

$$H(\omega) = H_{LSB}(\omega) \begin{cases} 1, & |\omega| < \omega_c \\ 0, & |\omega| \geqslant \omega_c \end{cases} \qquad (4.2.8)$$

The frequency spectrum of SSB is

$$S_{SSB}(\omega) = S_{DSB}(\omega) \cdot H(\omega) \qquad (4.2.9)$$

Figure 4. 2. 8 shows the process of upper sideband SSB modulation, using the filter method.

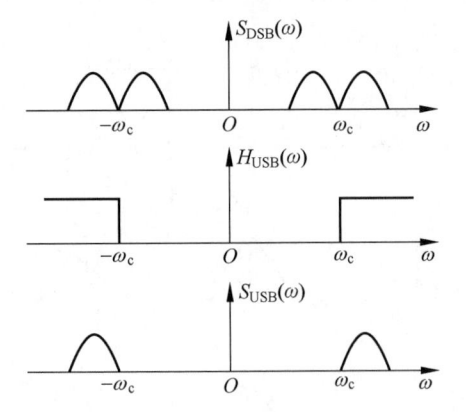

Figure 4. 2. 8   Spectrum of SSB signal containing the upper sideband

The message spectrum must have an energy gap centered at the origin. This requirement is naturally satisfied by voice signals, whose energy gap is about 600 Hz wide (from −300 Hz to 300 Hz), as shown in Figure 4.2.2.

The equations (4.2.7) and (4.2.8) are ideal filters, which are difficult to realize in practice. The above method is also called filter method（滤波法）to generate the SSB signal. Now let us discuss another method: phase-shifting method（相移法）(see Figure 4.2.9).

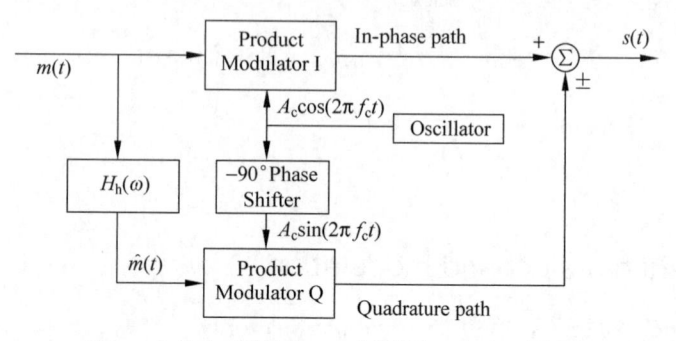

Figure 4.2.9 Modulator of SSB by phase-shifting

In this model, $H_h(\omega)$ is the transfer function of Hilbert transform（希尔伯特变换）, which is the $\pi/2$ phase shifting of input signal. We can get the time domain equation of SSB from Figure 4.2.9.

$$s_{SSB}(t) = \frac{1}{2}m(t)\cos\omega_c t \mp \frac{1}{2}\hat{m}(t)\sin\omega_c t \tag{4.2.10}$$

where $\hat{m}(t)$ is the Hilbert transform of $m(t)$.

At the receiver, the coherent detection is used to recovery the message signal. Carrier $\cos\omega_c t$ multiplying the received signal $[S_{USB}（上边带）$ or $S_{SSB}（下边带）]$ with the carrier is equivalent to the convolution（卷积）of the carrier frequency spectrum with the signal frequency spectrum in frequency domain. We consider upper-sideband signal as an example. $S_{SSB}(f)$ convolutes with $C(f)$, and then passes through a LPF, the result is the frequency Spectrum of $M(f)$, as illustrated in Figure 4.2.10.

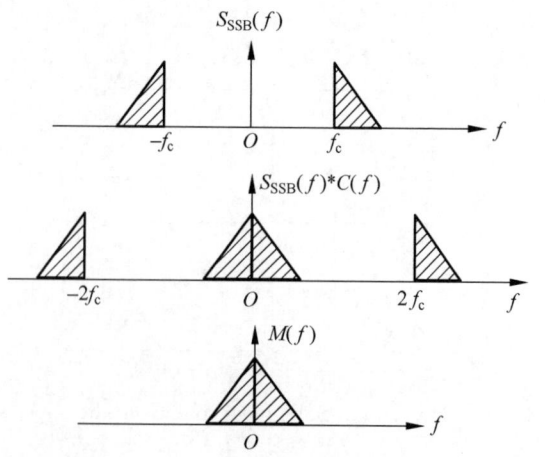

Figure 4.2.10 Demodulation of SSB

4.2.4　Vestigial sideband modulation

Vestigial sideband(VSB,残留边带)modulation: Only a vestige of one of the sidebands and corresponding modified version of the other sideband are transmitted. (One of the sidebands is partially suppressed and a vestige of the other sideband is transmitted to compensate for that suppression.)

We generate a DSB-SC modulated wave and then pass it through a band pass filter (BPF), which is specially designed to distinguish VSB modulation from SSB modulation. Frequency spectrum width of VSB signal is between that of DSB and SSB signals, and its frequency spectrum contains carrier components and very low frequency components. VSB modulation is currently used in analog TV broadcasting systems extensively.

How to determine the $H(f)$ of VSB modulation. Here we use the system analysis method to get the $H(f)$ from frequency domain, as shown in Figure 4.2.11.

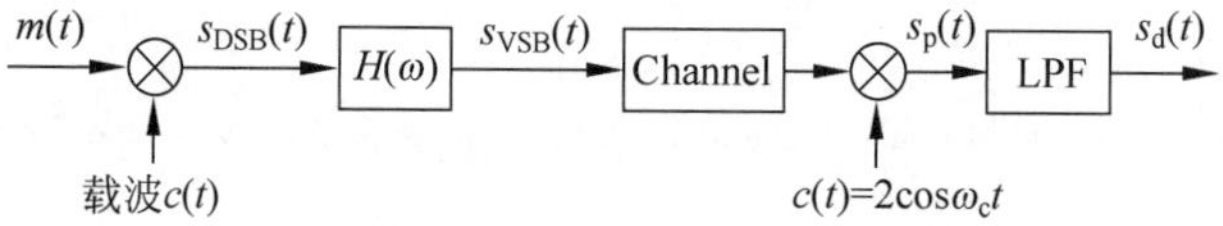

Figure 4.2.11　The VSB communication system

At the transmitter, frequency spectrum expression of the output signal of the multiplier is

$$S_{DSB} = \frac{1}{2}[M(\omega + \omega_c) + M(\omega - \omega_c)] \tag{4.2.11}$$

Frequency Spectrum of VSB should be

$$S_{VSB}(\omega) = S_{DSB}(\omega) \cdot H(\omega) = \frac{1}{2}[M(\omega + \omega_c) + M(\omega - \omega_c)]H(\omega) \tag{4.2.12}$$

In order to find the condition that $H(\omega)$ should be satisfied, analyze it from the demodulation, and neglect the effect of noise.

After multiplication of signal $s(t)$ and local carrier $\cos\omega_c t$, the frequency spectrum of $s_p(t)$ will be the result of the frequency shifting of $S_{VSB}(\omega)$:

$$S_p(\omega) = [S_{VSB}(\omega + \omega_c) + S_{VSB}(\omega - \omega_c)] \tag{4.2.13}$$

Substituting (4.2.12) into (4.2.13)

$$S_p(\omega) = \frac{1}{2}[M(\omega + 2\omega_c) + M(\omega)]H(\omega + \omega_c) + \frac{1}{2}[M(\omega) + M(\omega - 2\omega_c)]H(\omega - \omega_c) \tag{4.2.14}$$

The items $M(\omega + 2\omega_c)$ and $M(\omega - 2\omega_c)$ can be filtered out by LPF. Hence, the output of system is

$$S_d(\omega) = \frac{1}{2}M(\omega)[H(\omega + \omega_c) + H(\omega - \omega_c)] \tag{4.2.15}$$

For distortion-less transmission, we require the following condition of $H(\omega)$:

$$\boxed{H(\omega + \omega_c) + H(\omega - \omega_c) = C, \quad |\omega| \leqslant \omega_H} \tag{4.2.16}$$

Where ω_H is the cut-off angular frequency of message singal, where C is a constant.

The above equation is prerequisite（必要条件）for the filter characteristics to produce VSB signal. The cut-off characteristics of the filter is complementary symmetry（互补对称）with respect to the carrier frequency f_c（在 f_c 处具有互补对称性）(see Figure 4.2.12).

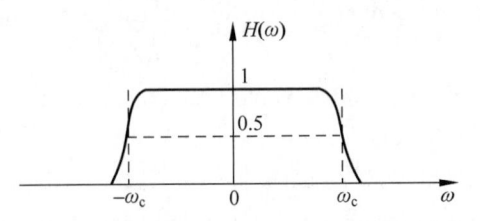

(a) Amplitude-frequency characteristic of residual partial upper band filter

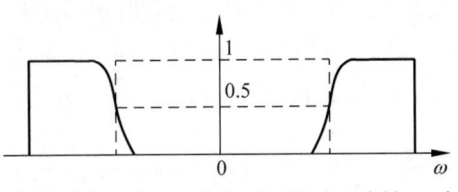

(b) Amplitude-frequency characteristic of residual partial lower band filter

Figure 4.2.12　The filter characteristic of VSB system

Accordingly, the transmission bandwidth of VSB modulation is

$$B_T = B + f_v \qquad (4.2.17)$$

where B is the message bandwidth, and f_v is the width of vestigial sideband.

The time domain of VSB modulated wave is described as

$$s(t) = \frac{1}{2}A_c m(t)\cos(2\pi f_c t) \pm \frac{1}{2}A_c m'(t)\sin(2\pi f_c t) \qquad (4.2.18)$$

where $m'(t)$ in the quadrature component of $s(t)$ and it is obtained by passing the message signal $m(t)$ through a filter whose frequency response $H_Q(f)$ satisfies the following requirement:

$$H_Q(f) = j[H(f - f_c) - H(f + f_c)], \quad \text{for} - f_H \leqslant f \leqslant f_H \qquad (4.2.19)$$

Figure 4.2.13 displays a plot of the frequency response $H_Q(f)$, scaled by $1/j$.

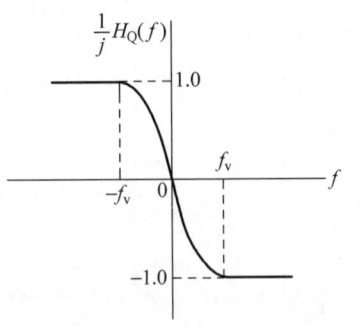

Figure 4.2.13　Frequency response of a filter for producing the quadrature component of the VSB modulated wave

It is of interest to note that SSB modulation may be viewed as a special case of VSB modulation. Specially, when the vestigial sideband is reduced to zero ($f_v=0$).

4.3 Anti-noise performance of linear demodulation

视频讲解

Figure 4.3.1 shows the noisy receiver model. The noise $n(t)$ is addictive white Gaussian noise (AWGN). The receiver consists of an ideal BPF following by an ideal demodulator. We also assume that BPF has a bandwidth equal to the transmission bandwidth B of the modulated signal $s(t)$ and the mid-band frequency equal to the carrier frequency f_c.

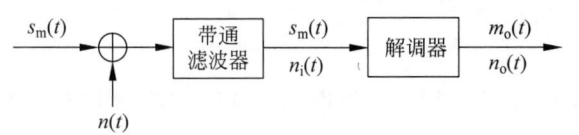

Figure 4.3.1 Receiver model

Suppose the midband frequency of the band-pass filter is f_c, then the PSD of noise after passing this filter is shown in Figure 4.3.2.

We may treat the filtered noise $n(t)$ as a narrowband noise represented in the following form:

$$n(t)=n_I(t)\cos(2\pi f_c t)-n_Q(t)\sin(2\pi f_c t) \tag{4.3.1}$$

The filtered signal $x(t)$ available for demodulation is defined by

$$x(t)=s(t)+n(t)$$

The details of $s(t)$ depend on the type of modulation used.

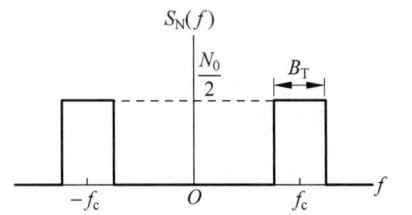

Figure 4.3.2 Idealized frequency response of the noise filtered by a BPF

The input signal-to noise ratio $(\mathrm{SNR})_I$ at the demodulator input is defined as:

$$(\mathrm{SNR})_I=\frac{\text{average power of } s_m(t)}{\text{average power of } n_i(t)} \tag{4.3.2}$$

The output signal-to-noise ratio $(\mathrm{SNR})_O$ at the receiver output is defined as:

$$(\mathrm{SNR})_O=\frac{\text{average power of the demodulated message signal } m_o(t)}{\text{average power of the noise } n_o(t)} \tag{4.3.3}$$

We define modulation gain (调制制度增益) G for the receiver as follows:

$$\boxed{G=\frac{(\mathrm{SNR})_O}{(\mathrm{SNR})_I}} \tag{4.3.4}$$

We give the definition of **the channel signal-to-noise ratio $(\mathrm{SNR})_C$** as the ratio of the average power of the modulated signal to the average power of channel noise in the message bandwidth both measured at the receiver input.

$$(SNR)_C = \frac{\text{The average power of the modulated message signal } s_m(t)}{\text{The average power of the channel noise } n_c(t) \text{in the message bandwidth}}$$

(4.3.5)

We define a Figure of merit for the receiver as follows:

$$\boxed{\text{Figure of merit} = \frac{(SNR)_O}{(SNR)_C}}$$

(4.3.6)

The higher the value of Figure of merit or G, the better will the anti-noise performance of the receiver be.

4.3.1　Noise in linear receiver using coherent detection

Figure 4.3.3 shows the DSB/SSB using a coherent detector.

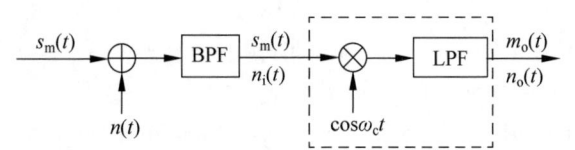

Figure 4.3.3　The model of DSB/SSB receiver using coherent detection

The comparison of anti-noise performance between DSB and SSB is as follows:

(1) DSB input signal of receiver is (modulated signal):

$$s_m(t) = m(t)\cos\omega_c t$$

(4.3.7)

$s_m(t)$ multiplies with carrier wave $\cos\omega_c t$

$$m(t)\cos^2\omega_c t = \frac{1}{2}m(t) + \frac{1}{2}m(t)\cos2\omega_c t$$

(4.3.8)

Then it passes through the LPF, the output signal is

$$m_o(t) = \frac{1}{2}m(t)$$

(4.3.9)

so

$$S_i = \overline{s_m^2(t)} = \overline{[m(t)\cos\omega_c t]^2} = \frac{1}{2}\overline{m^2(t)}$$

$$S_o = \overline{m_o^2(t)} = \frac{1}{4}\overline{m^2(t)}$$

The input noise of demodulator:

$$n_i(t) = n_c(t)\cos\omega_c t - n_s(t)\sin\omega_c t$$

(4.3.10)

Multiply with $\cos\omega_c t$:

$$n_i(t)\cos\omega_c t = [n_c(t)\cos\omega_c t - n_s(t)\sin\omega_c t]\cos\omega_c t$$

$$= \frac{1}{2}n_c(t) + \frac{1}{2}[n_c(t)\cos2\omega_c t - n_s(t)\sin2\omega_c t]$$

(4.3.11)

Then it passes through the LPF, the output noise is

$$n_o(t) = \frac{1}{2}n_c(t) \qquad (4.3.12)$$

so

$$N_o = \frac{1}{4}\overline{n_i^2(t)} = \frac{1}{4}N_i = \frac{1}{4}n_0 B$$

The input $(SNR)_I$ and **the channel signal-to-noise ratio** $(SNR)_C$ are

$$\frac{S_i}{N_i} = \frac{\frac{1}{2}\overline{m^2(t)}}{n_0 B} \quad \text{and} \quad (SNR)_C = \frac{\frac{1}{2}\overline{m^2(t)}}{\frac{1}{2}n_0 B} \qquad (4.3.13)$$

The output SNR is:

$$\frac{S_o}{N_o} = \frac{\frac{1}{4}\overline{m^2(t)}}{\frac{1}{4}N_i} = \frac{\overline{m^2(t)}}{n_0 B} \qquad (4.3.14)$$

So we can get:

$$G_{DSB} = \frac{S_o/N_o}{S_i/N_i} = 2$$

$$\text{Figure of merit} = \frac{(SNR)_O}{(SNR)_C}\bigg|_{DSB} = 1 \qquad (4.3.15)$$

(2) For the SSB system:

$$s_m(t) = \frac{1}{2}m(t)\cos\omega_c t \mp \frac{1}{2}\hat{m}(t)\sin\omega_c t \qquad (4.3.16)$$

When this SSB signal multiplies with $\cos\omega_c t$, and passes through the LPF, the output is

$$m_o(t) = \frac{1}{4}m(t)$$

so

$$S_o = \overline{m_o^2(t)} = \frac{1}{16}\overline{m^2(t)}$$

Because

$$S_i = \overline{s_m^2(t)} = \frac{1}{4}\overline{[m(t)\cos\omega_c t \mp \hat{m}(t)\sin\omega_c t]^2} = \frac{1}{4}\left[\frac{1}{2}\overline{m^2(t)} + \frac{1}{2}\overline{m^2(t)}\right]$$

and

$$\overline{m^2(t)} = \overline{\hat{m}^2(t)}$$

It is easy to get

$$S_i = \frac{1}{4}\overline{m^2(t)}$$

so

$$\frac{S_i}{N_i} = \frac{\frac{1}{4}\overline{m^2(t)}}{\frac{n_0}{2}B} = \frac{\overline{m^2(t)}}{2n_0 B} \quad \text{and} \quad (SNR)_C = \frac{\frac{1}{4}\overline{m^2(t)}}{\frac{n_0}{2}B} = \frac{\overline{m^2(t)}}{2n_0 B}$$

The output SNR is

$$\frac{S_o}{N_o} = \frac{\frac{1}{16}\overline{m^2(t)}}{\frac{1}{4}\frac{n_0}{2}B} = \frac{\overline{m^2(t)}}{2n_0 B}$$

Finally，we can get

$$G_{SSB} = \frac{S_o/N_o}{S_i/N_i} = 1 \quad \text{and} \quad \text{Figure of merit} = \frac{(SNR)_O}{(SNR)_C}\bigg|_{DSB} = 1 \qquad (4.3.17)$$

The differences between DSB and SSB demodulator are the central frequency and bandwidth of BPF.

> **Important conclusion：**
>
> For the same average transmitted (or modulated) signal power and the same average noise power in the message bandwidth，a coherent SSB receiver will have exactly the same output signal-to noise ratio as a coherent DSB-SC receiver.
>
> Because the figure of merit is equal to 1，that is to say neither DSB nor SSB demodulation offers high quality of reception. The anti-noise performance does not improve at the receivers.

4.3.2　Noise in AM receivers using envelope detection

The noise analysis of AM system is using an envelope detection，as shown in Figure 4.3.4.

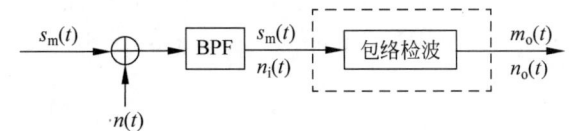

Figure 4.3.4　The model of AM receiver using envelope detection

The input signal is

$$s_m(t) = [A_0 + m(t)]\cos\omega_c t \qquad (4.3.18)$$

where $|m(t)|_{max} \leqslant A_0$.

The input noise $n_i(t)$ is：

$$n_i(t) = n_c(t)\cos\omega_c t - n_s(t)\sin\omega_c t \qquad (4.3.19)$$

We can get

$$S_i = \overline{s_m^2(t)} = \frac{A_0^2}{2} + \frac{\overline{m^2(t)}}{2}$$

$$N_i = \overline{n_i^2(t)} = n_0 B \qquad (4.3.20)$$

The input signal-to-noise ratio is

$$\frac{S_i}{N_i} = \frac{A_0^2 + \overline{m^2(t)}}{2n_0 B} \qquad (4.3.21)$$

The mixed signal $s_m(t) + n_i(t)$ is equal to

$$s_m(t) + n_i(t) = [A_0 + m(t) + n_c(t)]\cos\omega_c t - n_s(t)\sin\omega_c t$$

$$= E(t)\cos[\omega_c t + \psi(t)] \qquad (4.3.22)$$

where $E(t) = \sqrt{[A_0 + m(t) + n_c(t)]^2 + n_s^2(t)}$ and $\psi(t) = \arctan\left[\dfrac{n_s(t)}{A_0 + m(t) + n_c(t)}\right]$.

Obviously, $E(t)$ is the envelope, its expression is non-linear and complex. In order to simplify this problem, some approximation methods need to be used.

(1) High signal to noise ratio: $[A_0 + m(t)] \gg \sqrt{n_c^2(t) + n_s^2(t)}$

The envelope

$$E(t) = \sqrt{[A_0 + m(t)]^2 + 2[A_0 + m(t)]n_c(t) + n_c^2(t) + n_s^2(t)}$$

$$\approx \sqrt{[A_0 + m(t)]^2 + 2[A_0 + m(t)]n_c(t)}$$

$$\approx [A_0 + m(t)]\left[1 + \frac{2n_c(t)}{A_0 + m(t)}\right]^{1/2}$$

$$\approx [A_0 + m(t)]\left[1 + \frac{n_c(t)}{A_0 + m(t)}\right]$$

$$= A_0 + m(t) + n_c(t) \qquad (4.3.23)$$

Here the following approximate formula is used:

$$(1 + x)^{\frac{1}{2}} \approx 1 + \frac{x}{2}, \qquad |x| \ll 1 \qquad (4.3.24)$$

After the envelope detector, the output signal and noise are

$$S_o = \overline{m^2(t)}$$

$$N_o = \overline{n_c^2(t)} = \overline{n_i^2(t)} = n_0 B \qquad (4.3.25)$$

So

$$\frac{S_o}{N_o} = \frac{\overline{m^2(t)}}{n_0 B} \qquad (4.3.26)$$

$$G_{AM} = \frac{S_o/N_o}{S_i/N_i} = \frac{2\overline{m^2(t)}}{A_0^2 + \overline{m^2(t)}} \qquad (4.3.27)$$

For the 100% modulation $(|m(t)|_{max} = A_0)$

$$G_{AM} = \frac{2}{3} \quad \text{and} \quad \text{Figure of merit} = \frac{(\text{SNR})_O}{(\text{SNR})_C} = \frac{1}{3} \qquad (4.3.28)$$

This means that, assuming other factors are equal, an AM system (using envelope detection) must transmit three times as much average power as a suppressed-carrier system (using coherent detection) to achieve the same quality of noise performance.

(2) Low signal to noise ratio:

$$[A_0 + m(t)] \ll \sqrt{n_c^2(t) + n_s^2(t)}$$

The output of the envelop detector is

$$E(t) = R(t) + [A_0 + m(t)]\cos\theta(t) \qquad (4.3.29)$$

where $R(t) = \sqrt{n_c^2(t) + n_s^2(t)}$ and $\theta(t) = \arctan[n_s(t)/n_c(t)]$.

This relation reveals that when the signal-to-noise ratio is low, the detector output has no component strictly proportional to the message signal $m(t)$, as shown in Figure 4.3.5.

Threshold effects（门限效应）：Below a specific carrier-to-noise value (the threshold), the noise performance of a detector deteriorates much more rapidly than proportionally to the carrier-to-noise ratio.

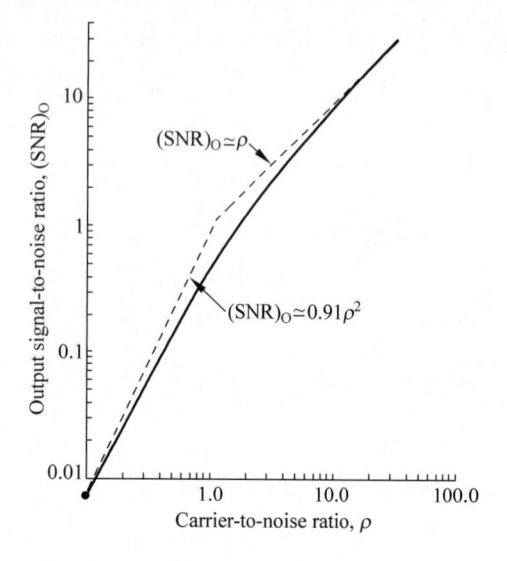

Figure 4.3.5　Output signal-to-noise ratio of an envelope detector for varying carrier-to-noise ratio

Example 4.3.1：the AM signal $s(t) = A_c[1 + k_a m(t)]\cos 2\pi f_c t$ is applied to the system shown in Figure 4.3.6. Assuming that $|k_a m(t)| < 1$ for t and the message signal $m(t)$ is limited to the interval $-W \leqslant f \leqslant W$ and that the carrier frequency $f_c > 2W$ show that can be obtained from the square-rooter output v_3.

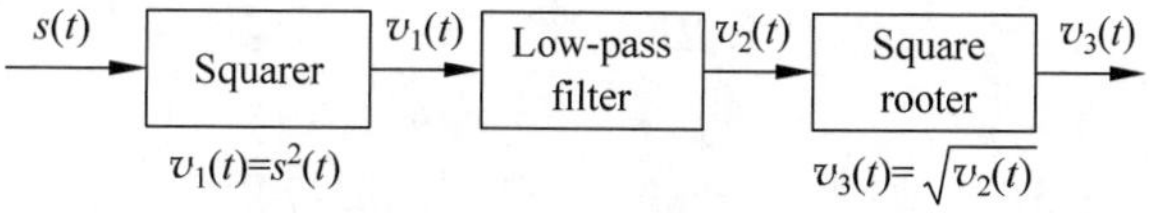

$$s(t) \rightarrow \boxed{\text{Squarer}} \xrightarrow{v_1(t)} \boxed{\begin{array}{c}\text{Low-pass}\\\text{filter}\end{array}} \xrightarrow{v_2(t)} \boxed{\begin{array}{c}\text{Square}\\\text{rooter}\end{array}} \xrightarrow{v_3(t)}$$

$$v_1(t) = s^2(t) \qquad\qquad\qquad v_3(t) = \sqrt{v_2(t)}$$

Figure 4.3.6　AM demodulator

Solution：The squarer output is

$$v_1(t) = A_c^2[1 + k_a m(t)]^2 \cos^2 2\pi f_c t$$
$$= \frac{A_c^2}{2}[1 + 2k_a m(t) + m^2(t)][1 + \cos 4\pi f_c t]$$

The output of the Low-pass filter is

$$v_2(t) = \frac{A_c^2}{2}[1 + k_a m(t)]^2$$

So the output of square rooter is

$$v_3 = \frac{A_c}{\sqrt{2}}[1 + k_a m(t)]$$

Homework（part 1）

4.1　The following Figure 4.1 shows the circuit diagram of a square-law modulator. The signal applied to the nonlinear device is relatively weak, such that it can be represented by a square law:

$$v_2(t) = a_1 v_1(t) + a_2 v_1^2(t)$$

where a_1 and a_2 are constants, $v_1(t)$ is the input voltage, and $v_2(t)$ is the output voltage. The input voltage is defined by $v_1(t) = A_c \cos 2\pi f_c t + m(t)$, where $m(t)$ is a message signal and $A_c \cos 2\pi f_c t$ is the carrier wave.

（1）Evaluate the output voltage $v_2(t)$;

（2）Specify the frequency response that the tuned circuit in the following figure must satisfy in order to generate an AM signal with f_c as the carrier frequency;

（3）What is the amplitude sensitivity of this AM signal.

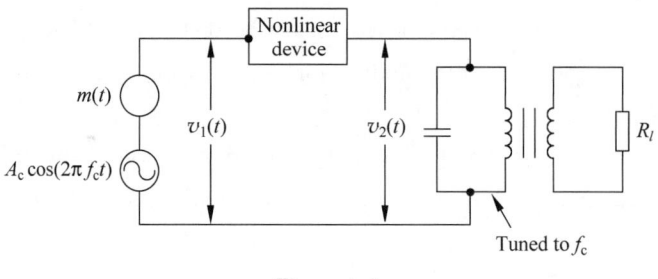

Figure 4.1

4.2　Consider a message signal $m(t)$ with the spectrum shown in Figure 4.2. The message bandwidth is $W = 1\text{kHz}$. This signal is applied to a product modulator, together with a carrier wave $A_c \cos 2\pi f_c t$, producing the DSB-SC modulated signal $s(t)$. The modulated signal is next applied to a coherent detector, determine the spectrum of the detector output when:

（1）the carrier frequency $f_c = 1.25\text{kHz}$;

（2）the carrier frequency $f_c = 0.75\text{kHz}$.

What is the lowest carrier frequency for which each component of the modulated signal $s(t)$ is uniquely determined by $m(t)$.

4.3　The waveform of message signal is shown in Figure 4.3. Draw the waveforms of DSB and AM signals and compare the difference after the envelope detector.

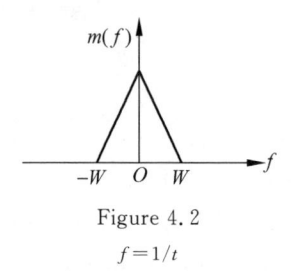

Figure 4.2

$f = 1/t$

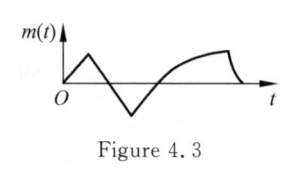

Figure 4.3

4. 4　　Suppose the noise double-side PSD of a channel is $P_n(f) = 0.5 \times 10^{-3}$ W/Hz, on this channel the DSB or SSB(upper side-band) or AM(The side-band power is 10kW, the power of carrier wave is 40kW)is transmitted. The bandwidth of message signal $m(t)$ is in 5kHz, and the frequency of carrier wave is 100kHz. The power of modulated signal is 10kW. If there is a band-pass filter before receiver, try to find:

（1）the mid-band frequency（中心频率）of this ideal band-pass filter for DSB/SSB signals;

（2）the input SNR of demodulator;

（3）the output SNR of demodulator, and draw the output noise PSD of DSB demodulator;

（4）the modulation system gain G of AM demodulator.

视频讲解

4.4　Angle modulation（non-linear modulation process）

The angle of the carrier wave is varied according to the baseband（message）signal in angle modulation method. In this modulation, the amplitude of the carrier wave is maintained a constant.

An important feature of angle modulation is that it can provide better anti-noise performance than linear modulation by exchanging channel bandwidth（以带宽换取高信噪比）.

4.4.1　Basic definitions

The expression of angle modulation is

$$s_m(t) = A\cos[\omega_c t + \varphi(t)] \tag{4.4.1}$$

where $\varphi(t)$ is a function of $m(t)$.

When $\varphi(t) = K_p m(t)$

$$s_{PM}(t) = A\cos[\omega_c t + K_p m(t)]$$

—Phase modulation

When $\dfrac{\mathrm{d}\varphi(t)}{\mathrm{d}t} = K_f m(t)$ 　or　 $\varphi(t) = K_f \displaystyle\int m(\tau)\mathrm{d}\tau$

$$s_{FM}(t) = A\cos\left[\omega_c t + K_f \int m(\tau)\mathrm{d}\tau\right]$$

—Frequency modulation

Instantaneous phase（瞬时相位）：$\omega_c t + \varphi(t)$.

Instantaneous phase deviation（瞬时相位偏移）：$\varphi(t)$.

Instantaneous frequency（瞬时频率）：$\mathrm{d}[\omega_c t + \varphi(t)]/\mathrm{d}t$.

Instantaneous frequency deviation（瞬时频率偏移）：$\mathrm{d}\varphi(t)/\mathrm{d}t$.

The waveform of angle modulation is shown in Figure 4.4.1. As can be seen, if there is no information about the message signal, it is difficult to distinguish FM and PM signal.

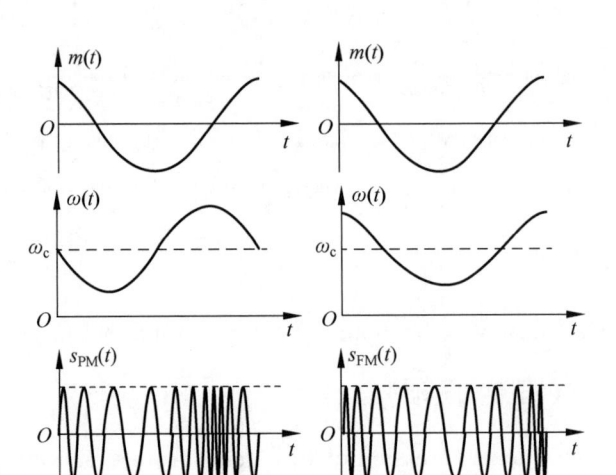

Figure 4.4.1   Waveform of angle modulation

Now we consider a cosine modulated signal defined by

$$m(t) = A_m \cos\omega_m t$$

$$s_{PM}(t) = A\cos[\omega_c t + K_p A_m \cos\omega_m t] = A\cos[\omega_c t + m_p \cos\omega_m t] \qquad (4.4.2)$$

where $m_p = K_p A_m$.

m_p is called the modulation index of PM signal, representing the maximum of phase deviation.

The expression of FM is as the following equation:

$$s_{FM}(t) = A\cos\left[\omega_c t + K_f \int m(\tau)d\tau\right]$$

$$= A\cos\left[\omega_c t + K_f A_m \int \cos\omega_m \tau d\tau\right]$$

$$= A\cos(\omega_c t + m_f \sin\omega_m t) \qquad (4.4.3)$$

where $m_f = K_f A_m = \dfrac{\Delta\omega}{\omega_m} = \dfrac{\Delta f}{f_m} = \beta$, $\Delta\omega = K_f A_m$ is the largest angular frequency deviation（最大角频偏）. $\Delta f = m_f f_m$ is the largest frequency deviation（最大频偏）.

m_f is commonly called the modulation index of FM signal.

Comparing the equations of PM and FM signals. We can find that an FM signal may be regarded as a PM signal in which the modulating wave is $\int m(\tau)d\tau$ in place of $m(t)$. This means that an FM signal can be generated by first integrating $m(t)$ and then using the result as the input to a phase modulator. Conversely, a PM signal can be generated by first differentiating $m(t)$ and then using the result as the input to a frequency modulator (see Figure 4.4.2).

Depending on the value of the modulation index m_f (or β), we may distinguish two cases:

(1) Narrowband FM (NBFM), for which β is small compared to one radian.

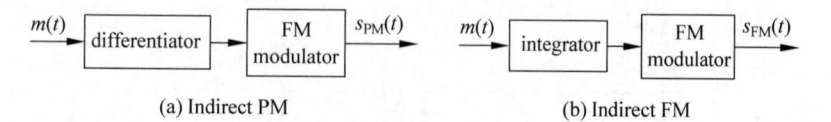

(a) Indirect PM   (b) Indirect FM

Figure 4. 4. 2   The indirect generation of PM and FM

$$\left(\left|K_f\int_{-\infty}^{t}m(\tau)d\tau\right|\ll\frac{\pi}{6}\right)$$

(2) Wideband FM(WBFM), for which β is large compared to one radian.

4.4.2   NBFM(窄带调频)

We expand the definition expression of FM signal from equation(4. 4. 3):

$$s_{FM}(t)=A\cos\left[\omega_c t+K_f\int_{-\infty}^{t}m(\tau)d\tau\right]$$

$$=A\cos\omega_c t\cos\left[K_f\int_{-\infty}^{t}m(\tau)d\tau\right]-A\sin\omega_c t\sin\left[K_f\int_{-\infty}^{t}m(\tau)d\tau\right] \quad (4.4.4)$$

Assuming that m_f is small compared to one radian, we may use the following approximations:

$$\cos\left[K_f\int_{-\infty}^{t}m(\tau)d\tau\right]\approx 1$$

$$\sin\left[K_f\int_{-\infty}^{t}m(\tau)d\tau\right]\approx K_f\int_{-\infty}^{t}m(\tau)d\tau$$

$$s_{NBFM}(t)\approx A\cos\omega_c t-\left[AK_f\int_{-\infty}^{t}m(\tau)d\tau\right]\sin\omega_c t \quad (4.4.5)$$

According to the following Fourier transforms:

$$m(t)\Leftrightarrow M(\omega)$$

$$\cos\omega_c t\Leftrightarrow\pi[\delta(\omega+\omega_c)+\delta(\omega-\omega_c)]$$

$$\sin\omega_c t\Leftrightarrow j\pi[\delta(\omega+\omega_c)-\delta(\omega-\omega_c)]$$

$$\int m(t)dt\Leftrightarrow\frac{M(\omega)}{j\omega}$$

$$\left[\int m(t)dt\right]\sin\omega_c t\Leftrightarrow\frac{1}{2}\left[\frac{M(\omega+\omega_c)}{\omega+\omega_c}-\frac{M(\omega-\omega_c)}{\omega-\omega_c}\right] \quad (4.4.6)$$

The frequency domain expression of NBFM is

$$s_{NBFM}(\omega)=\pi A[\delta(\omega+\omega_c)+\delta(\omega-\omega_c)]+\frac{AK_f}{2}\left[\frac{M(\omega-\omega_c)}{\omega-\omega_c}-\frac{M(\omega+\omega_c)}{\omega+\omega_c}\right] \quad (4.4.7)$$

Compare the spectrum of AM with the above equation,

$$S_{AM}(\omega)=\pi A[\delta(\omega+\omega_c)+\delta(\omega-\omega_c)]+\frac{1}{2}[M(\omega+\omega_c)+M(\omega-\omega_c)] \quad (4.4.8)$$

Suppose the message signal is $m(t)=A_m\cos\omega_m t$, we can find the similarities and differences between AM and NBFM from the above equations and the vector graphs in Figure 4. 4. 3:

(1) Both of them have two sidebands, and they have the same bandwidth. But one of NBFM sideband is reversed phase compared with the sideband of AM signal.

(2) There are two frequency weighting factors $\dfrac{1}{\omega+\omega_c}$ and $\dfrac{1}{\omega-\omega_c}$ of the two sidebands of NBFM, which cause the change of spectrum, so it is a nonlinear modulation.

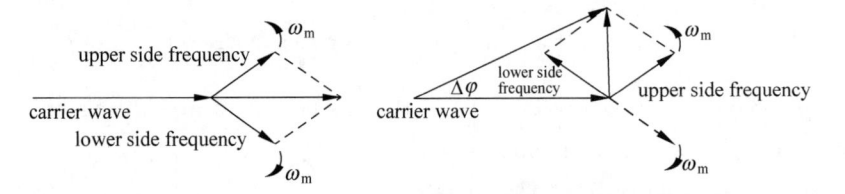

Figure 4.4.3   The vector graph of AM and NBFM

4.4.3   Wide-band frequency modulation（WBFM）

Take the single-tone modulation for example:

$$m(t)=A_m\cos\omega_m t=A_m\cos2\pi f_m t \tag{4.4.9}$$

$$s_{FM}(t)=A\cos\omega_c t \cdot \cos(m_f\sin\omega_m t)-A\sin\omega_c t \cdot \sin(m_f\sin\omega_m t) \tag{4.4.10}$$

We may expand $s_{FM}(t)$ in the form of a Fourier series（傅里叶级数）

$$\cos(m_f\sin\omega_m t)=J_0(m_f)+\sum_{n=1}^{\infty}2J_{2n}(m_f)\cos2n\omega_m t$$

$$\sin(m_f\sin\omega_m t)=2\sum_{n=1}^{\infty}J_{2n-1}(m_f)\sin(2n-1)\omega_m t$$

where $J_n(m_f)$ is the nth order Bessel function（n 阶贝塞尔函数）of the first kind（第一类）.

Bessel function $J_n(m_f)$ has the following properties:

(1) $J_{-n}(m_f)=-J_n(m_f)$, n is odd

  $J_{-n}(m_f)=J_n(m_f)$, n is even

(2) For small values of the modulation index m_f, we have

$$\begin{cases} J_0(m_f)\approx 1 \\ J_1(m_f)\approx m_f/2 \\ J_n(m_f)\approx 0, \quad n>2 \end{cases}$$

(3) $\displaystyle\sum_{n=-\infty}^{\infty}J_n^2(m_f)=1$

Accordingly, we may get the following expansion of an FM signal:

$$s_{FM}(t)=A\sum_{n=-\infty}^{\infty}J_n(m_f)\cos(\omega_c+n\omega_m)t \tag{4.4.11}$$

The Fourier transform of FM signal is:

$$S_{FM}(\omega)=\pi A\sum_{-\infty}^{\infty}J_n(m_f)[\delta(\omega-\omega_c-n\omega_m)+\delta(\omega+\omega_c+n\omega_m)] \tag{4.4.12}$$

The one sideband (positive frequencies) spectrum is plotted in Figure 4.4.4.

The spectrum of FM signal contains a carrier component and infinite set of side frequencies located symmetrically on either side of the carrier at frequency separations of f_m, $2f_m$, $3f_m$, $\cdots$.

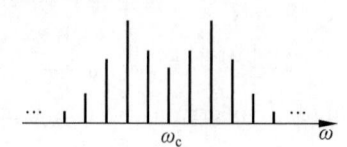

Figure 4. 4. 4　One sideband spectrum of WBFM

From the frequency domain，FM is not the frequency linear shifting of the message signal，so it belongs to nonlinear modulation.

1. Transmission bandwidth of FM signals

In theory，an FM signal contains infinite number of side frequencies. In practice，we find that the FM signal is effectively limited to a finite number of significant side frequencies $f_c，\pm f_m$. We may therefore specify an effective bandwidth required for the transmission of FM signal.

The effective bandwidth is

$$B_{FM} = 2(m_f + 1)f_m = 2(\Delta f + f_m) \qquad (4.4.13)$$

where m_f is the modulation index，Δf is the frequency deviation.

When $m_f \ll 1$，　$B_{FM} \approx 2f_m$　（NBFM）

When $m_f \gg 1$　$B_{FM} \approx 2\Delta f$　（WBFM）

The equation(4. 4. 13) relation is known as **Carson's rule**(卡森公式).

2. Average power of FM

According to $P_{FM} = \overline{s_{FM}^2(t)} = \dfrac{A^2}{2} \sum\limits_{n=-\infty}^{\infty} J_n^2(m_f)$ and $\sum\limits_{n=-\infty}^{\infty} J_n^2(m_f) = 1$，the power of FM is：

$$P_{FM} = \frac{A^2}{2} = P_c \qquad (4.4.14)$$

The physical explanation for this property is that the envelope of an FM signal is constant，so the average power of such a signal dissipated across l-ohm resistor is also constant.

4.4.4　Generating an FM signal

1. Direct FM

The carrier frequency is directly varied in accordance with the input baseband signal，which is readily accomplished using a voltage-controlled oscillator (VCO：压控振荡器)，as shown in Figure 4. 4. 5.

$$m(t) \longrightarrow \boxed{\text{VCO}} \longrightarrow s_{FM}(t)$$

Figure 4. 4. 5　Direct method of generating FM

2. Indirect FM

The indirect method of generating a wideband FM signal is illustrated in Figure 4. 4. 6. The modulated signal is firstly used to produce a NBFM signal，and frequency

乘风破浪

水木书荟

清华大学出版社
TSINGHUA UNIVERSITY PRESS

May all your wishes
come true

扬帆起航

水木书荟

清华大学出版社
TSINGHUA UNIVERSITY PRESS

如果知识是通向未来的大门，
我们愿意为你打造一把打开这扇门的钥匙！

https://www.shuimushuhui.com/

May all your wishes
come true

multiplications is next used to increase the frequency deviation to the desired level. The NBFM signal is generated by the integrator （积分器） and NBPM. The crystal control provides frequency stability. A multiplier is used to produce the desired WBFM signal.

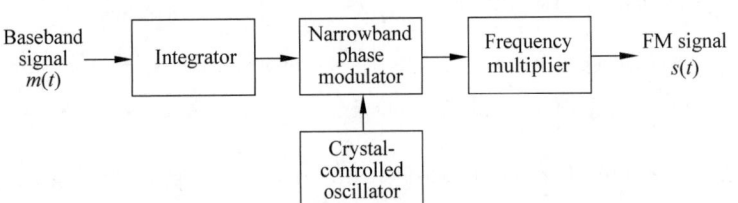

Figure 4.4.6   Block diagram of the indirect method of generating a wideband FM signal

A frequency multiplier as shown in Figure 4.4.7，consists of a nonlinear device followed by a band-pass filter. The input-output relation may be expressed in the following general form：

$$v(t) = a_1 s(t) + a_2 s^2(t) + \cdots + a_n s^n(t) \tag{4.4.15}$$

The input signal $s(t)$ is an FM signal defined by：

$$s(t) = A_c \cos\left[2\pi f_c t + 2\pi k_f \int_0^t m(\tau)\mathrm{d}\tau\right]$$

The output is：

$$s'(t) = A_c \cos\left[2\pi n f_c t + 2\pi n k_f \int_0^t m(\tau)\mathrm{d}\tau\right]$$

and its instantaneous frequency is

$$f_i'(t) = n f_c t + n K_f m(t)$$

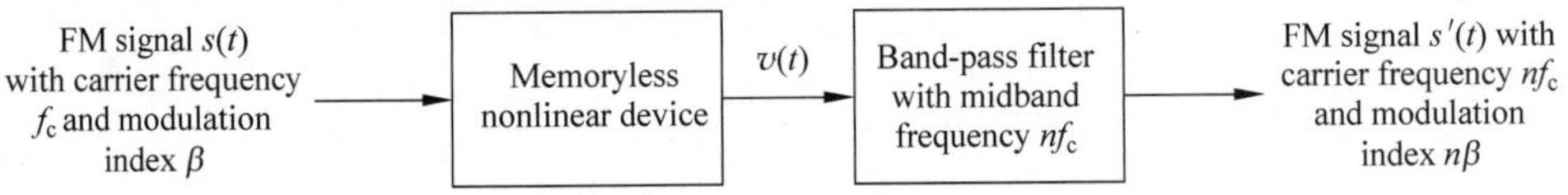

| FM signal $s(t)$ with carrier frequency f_c and modulation index β | → | Memoryless nonlinear device | $v(t)$ → | Band-pass filter with midband frequency nf_c | → | FM signal $s'(t)$ with carrier frequency nf_c and modulation index $n\beta$ |

Figure 4.4.7   Block diagram of frequency multiplier

The nonlinear processing circuit acts as a frequency multiplier （倍频器）.

4.4.5   Demodulation of FM signals

1. Non-coherent demodulation （NBFM，WBFM）

Use the frequency discriminator （鉴频器）：whose instantaneous output amplitude is directly proportional to the instantaneous frequency of the input FM signal.

$$s_{\mathrm{FM}}(t) = A \cos\left[\omega_c t + K_f \int_{-\infty}^t m(\tau)\mathrm{d}\tau\right] \tag{4.4.16}$$

The output of demodulator：

$$m_o(t) \propto K_f m(t) \tag{4.4.17}$$

The principle of amplitude frequency discriminator is as shown in Figure 4.4.8.

The output of differentiator is

$$s_d(t) = -A\left[\omega_c + K_f m(t)\right]\sin\left[\omega_c t + K_f \int_{-\infty}^t m(\tau)\mathrm{d}\tau\right] \tag{4.4.18}$$

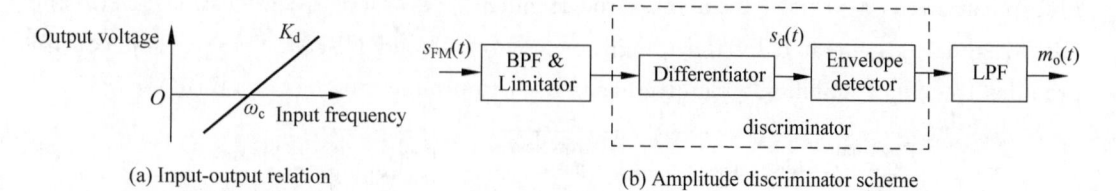

(a) Input-output relation (b) Amplitude discriminator scheme

Figure 4.4.8   Principle of amplitude discriminator

Then detect the envelope, filter out the DC component and pass through the LPF.

$$m_o(t) = K_d K_f m(t) \tag{4.4.19}$$

2. Coherent demodulation (NBFM)

The sketch map of coherent demodulation of NBFM is illustrated in Figure 4.4.9.

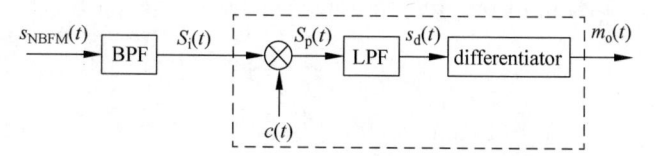

Figure 4.4.9   Coherent demodulation for NBFM

$$s_{NBFM}(t) = A\cos\omega_c t - A\left[K_f\int_{-\infty}^{t} m(\tau)d\tau\right] \cdot \sin\omega_c t \tag{4.4.20}$$

Suppose the carrier wave is

$$c(t) = -\sin\omega_c t$$

$$s_p(t) = -\frac{A}{2}\sin 2\omega_c t + \frac{A}{2}\left[K_f\int_{-\infty}^{t} m(\tau)d\tau\right] \cdot (1 - \cos 2\omega_c t) \tag{4.4.21}$$

After the low-pass filter

$$s_d(t) = \frac{A}{2}K_f\int_{-\infty}^{t} m(\tau)d\tau \tag{4.4.22}$$

The output of differentiator is

$$m_o(t) = \frac{AK_f}{2}m(t) \tag{4.4.23}$$

4.4.6   Noise in FM receivers

The discriminator consists of two components:

(1) A slope network or differentiator;

(2) An envelope detector, as shown in Figure 4.4.10.

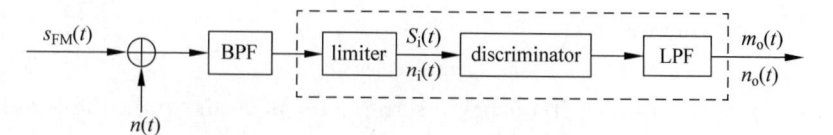

Figure 4.4.10   Model of FM receiver

1. Input signal-to-noise ratio

$$s_{FM}(t) = A\cos\left[\omega_c t + K_F \int_{-\infty}^{t} m(\tau)\,d\tau\right]$$

$$S_i = A^2/2$$

The input power of noise is:

$$N_i = n_0 B_{FM}$$

So the input signal to noise ratio is:

$$\frac{S_i}{N_i} = \frac{A^2}{2n_0 B_{FM}}$$

2. The output signal-to-noise ratio

First, we assume that the signal-to-noise ratio measured at the discriminator input is large compared with unity.

Because the signal and noise are independent.

Suppose $n = 0$, the output signal of modulator is

$$m_o(t) = K_d K_f m(t) \tag{4.4.24}$$

$$S_o = \overline{m_o^2(t)} = (K_d K_f)^2 \overline{m^2(t)} \tag{4.4.25}$$

Suppose $m(t) = 0$, the input of demodulator are the carrier-wave and narrowband Gaussian noise, that is

$$
\begin{aligned}
A\cos\omega_c t + n_i(t) &= A\cos\omega_c t + n_c(t)\cos\omega_c t - n_s(t)\sin\omega_c t \\
&= [A + n_c(t)]\cos\omega_c t - n_s(t)\sin\omega_c t \\
&= A(t)\cos[\omega_c t + \psi(t)]
\end{aligned}
\tag{4.4.26}
$$

where $A(t) = \sqrt{[A + n_c(t)]^2 + n_s^2(t)}$ and $\psi(t) = \arctan\dfrac{n_s(t)}{A + n_c(t)}$.

For the large SNR: $A \gg n_c(t)$ and $A \gg n_s(t)$, so the phase function is

$$\varphi(t) = \arctan\frac{n_s(t)}{A + n_c(t)} \doteq \arctan\frac{n_s(t)}{A} \tag{4.4.27}$$

When $x \ll 1$, $\arctan x = x$, $\varphi(t)$ can be simplified as:

$$\varphi(t) = \frac{n_s(t)}{A} \tag{4.4.28}$$

So the output of the discriminator (see Figure 4.4.11) is:

$$n_d(t) = k_d \frac{d\varphi(t)}{dt} = \frac{k_d}{A}\frac{dn_s(t)}{dt} \tag{4.4.29}$$

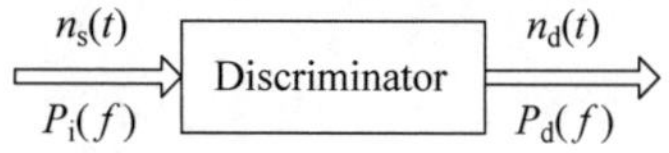

Figure 4.4.11　The input and output parameters of the discriminator

The transfer function of discriminator is the transmission characteristic of ideal differentiator.

$$|H(f)|^2 = |j2\pi f|^2 = (2\pi)^2 f^2 \tag{4.4.30}$$

$$P_{\mathrm{d}}(f) = \left(\frac{k_{\mathrm{d}}}{A}\right)^2 |H(f)|^2 \cdot P_{\mathrm{i}}(f) = \left(\frac{k_{\mathrm{d}}}{A}\right)^2 (2\pi)^2 f^2 n_0, \quad |f| < \frac{B_{\mathrm{FM}}}{2} \qquad (4.4.31)$$

The output noise power is

$$N_0 = \int_{-f_{\mathrm{m}}}^{f_{\mathrm{m}}} P_{\mathrm{d}}(f)\mathrm{d}f = \frac{8\pi^2 \cdot k_{\mathrm{d}}^2 \cdot n_0 f_{\mathrm{m}}^3}{3A^2} \qquad (4.4.32)$$

The output SNR is

$$\frac{S_0}{N_0} = \frac{3A^2 k_{\mathrm{f}}^2 \overline{m^2(t)}}{8\pi^2 n_0 f_{\mathrm{m}}^3} \qquad (4.4.33)$$

Suppose $m(t) = \cos\omega_{\mathrm{m}}t$ (single-tone demodulation), the modulation system gain is

$$G_{\mathrm{FM}} = \frac{S_o/N_o}{S_{\mathrm{i}}/N_{\mathrm{i}}} = \frac{3}{2}m_{\mathrm{f}}^2 \frac{B_{\mathrm{FM}}}{f_{\mathrm{m}}} \qquad (4.4.34)$$

For WBFM system, the anti-noise performance is:

$$G_{\mathrm{FM}} = 3m_{\mathrm{f}}^2(m_{\mathrm{f}} + 1) \qquad (4.4.35)$$

When $m_{\mathrm{f}} \gg 1$:

$$G_{\mathrm{FM}} = 3m_{\mathrm{f}}^3 \qquad (4.4.36)$$

Second, when the signal-to-noise ratio is small in FM receiver, there is a phenomenon called **threshold effect**, as can ben seen in Figure 4.4.12.

When the input signal to noise ratio is below a certain threshold, the output of dcmodulator decreases significantly（即当$\frac{S_{\mathrm{i}}}{N_{\mathrm{i}}}$低于某个门限时，解调后输出的$\frac{S_{\mathrm{o}}}{N_{\mathrm{o}}}$急剧恶化）.

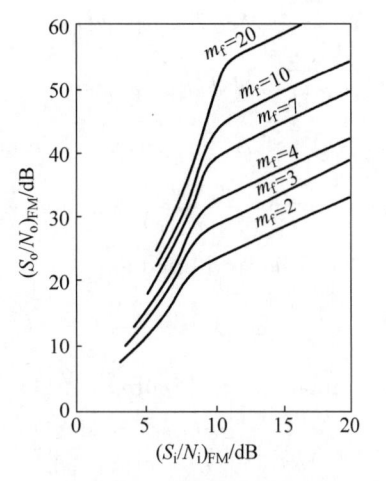

Figure 4.4.12   Dependence of output SNR on input SNR for FM receiver

4.5 Frequency-division multiplexing（频分复用）

The technique of separating the signals in frequency is referred to as frequency-division multiplexing (FDM), whereas the technique of separating the signals in time is called time-division multiplexing (TDM). In this section, we discuss the FDM systems, the block diagram is shown in Figure 4.5.1, and TDM systems will be discussed in next chapter.

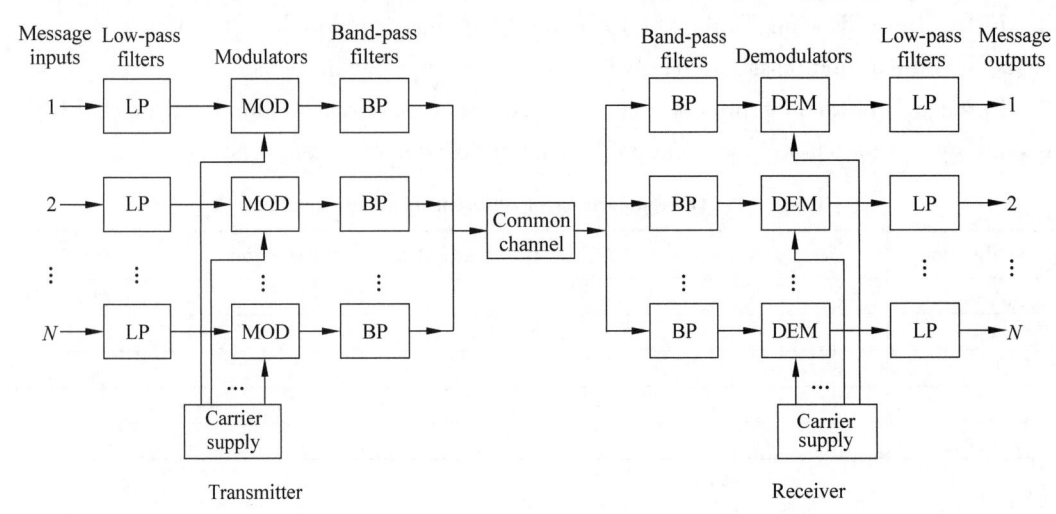

Figure 4.5.1   Block diagram of the FDM system

Summary and discussion

Continuous wave modulation is briefly studied in this chapter. This analog form modulation uses a sinusoidal carrier whose amplitude or angle is varied in accordance with a message signal. It is classified in two categories: linear modulation and nonlinear modulation.

Linear modulation is also called amplitude modulation, which the frequency is the displacement of the modulated signal (AM&DSB), or the filtering out of unnecessary frequency components after displacement (SSB&VSB). The bandwidth range is from B to 2B according to the different types.

Nonlinear modulation is also called angle modulation. It includes the FM and PM, which their frequencies are quite different from the modulated signal, appearing new frequency components. Hence, the bandwidth of the modulated signal increases considerably. The Carson's rule is used to calculate the bandwidth:

$$B_{FM} = 2(m_f + 1)f_m = 2(\Delta f + f_m)$$

Modulation system gain G is defined to give the anti-noise comparison of different modulations. Angular modulated signal has stronger anti-jamming ability at the expense of an excessive transmission bandwidth, compared with the linear modulation, and is especially suitable for transmission in the fading channel.

The model of each modulation and demodulation should be understood and remembered, including the coherent and non-coherent demodulations, especially the differences of AM, DSB, SSB and VSB, the comparisons between AM and NBFM.

Threshold effects occur in AM and FM non-coherent demodulation systems. Try to understand it.

Fill in the following Table 4.1 to understand these modulations from time-domain, frequency domain expressions, bandwidth and anti-noise performance.

Based on the analog modulation principle, frequency-division multiplexing（FDM）technology is adopted widely in the multi-channel carrier telephone systems.

Table 4.1   The comparisons of different modulation methods

Modulation	$s_m(t)$	$S_m(f)$	Bandwidth	Detection	G
AM					
DSB					
SSB					
VSB					
NBFM					

Homework（part 2）

4.5   Let the expression of an angular modulated signal be
$$s(t) = 10\cos[2 \times 10^6 \pi t + 10\cos 2000\pi t]$$
Find：（1）the maximum frequency deviation of the modulated signal.

（2）the maximum phase deviation of the modulated signal.

（3）the bandwidth of the modulated signal.

Terminologies

convolution	卷积	FDM	频分复用
complementary symmetry	互补对称	TDM	时分复用
modulation	调制	AM	调幅
demodulation	解调	DSB	双边带调制
modulation system gain	调制系统增益	SSB	单边带调制
reversals	反相	VSB	残留边带调制
Carson's rule	卡森公式	VCO	压控振荡器
synchronized	同步	FM	调频
threshold effect	门限效应	PM	调相
carrier wave	载波	NBFM	窄带调频
modulated signal	已调信号	WBFM	宽带调频

Chapter 5

Pulse modulation

Mind map:

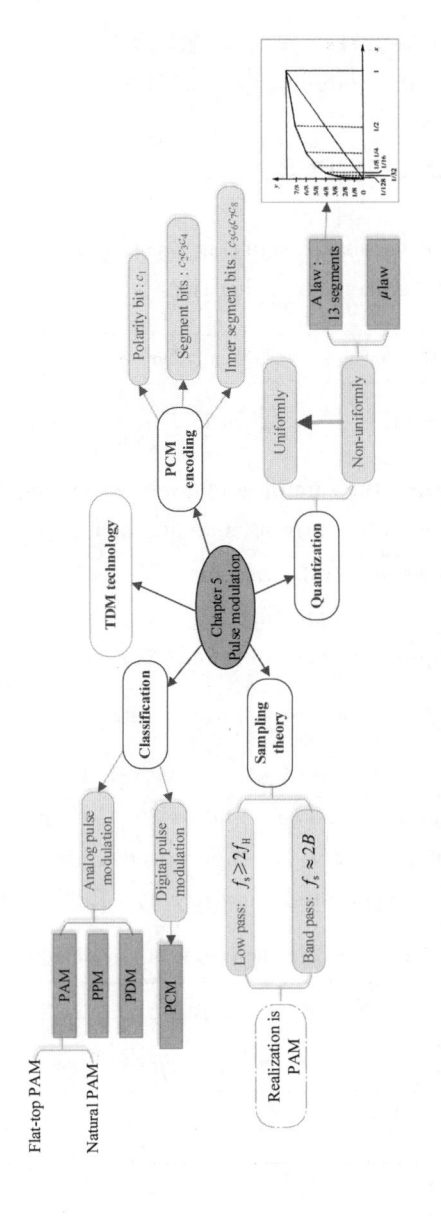

Mind map for
Chapter 5

In Chapter 4，some parameters of a sinusoidal carrier wave are varied continuously in accordance with the message signal. In this chapter，some parameters of a pulse train are varied in accordance with the message signal. The carrier wave is a pulse，which is different from CW modulation.

The following issues will be discussed：

- Sampling，which is basic to all forms of pulse modulation.
- Pulse amplitude modulation（PAM,脉冲幅度调制）（In discrete amplitude form）.
- Quantization（量化），after the process of sampling，in discrete form in both amplitude and time.
- Pulse-code modulation（PCM,脉冲编码调制），which is the standard method for the transmission of an analog message signal by digital means.
- TDM（时分复用）.
- Delta modulation（DM/ΔM,Δ调制/增量调制）.
  Differential pulse-code modulation（DPCM,差分脉冲编码调制）.

视频讲解

5.1 Sampling process

The process of digitizing an analog signal is illustrated in Figure 5.1.1. There are three main steps in the process of digitizing an analog signal：sampling，quantization and encoding.

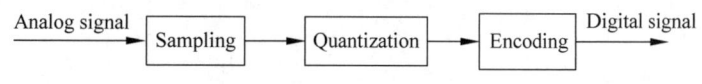

Figure 5.1.1 The process of digitizing an analog signal

1. The sampling theorem（the sampling of low-pass analog signals）

If the highest frequency of a continuous analog signal $s(t)$ is less than f_H，and if it is sampled by periodic impulses with interval time $T \leqslant 1/2f_H$，then $s(t)$ can be completely decided by these samples，as shown in Figure 5.1.2.

The ideal form of sampling process is called instantaneous sampling（瞬时采样）. T_s is the sampling period，and $f_s = \dfrac{1}{T_s}$ is the sampling rate.

The derived process is in the following Table 5.1.1：let the spectra of $m(t)$, $\delta_T(t)$ and $m_s(t)$ be expressed by $M(f), \Delta_\Omega(f)$ and $M_s(f)$. If the sampling frequency is lower than $2f_H$，the adjacent spectra will be superposed（重叠），hence the original signal spectrum $M(f)$ could not be separated correctly by LPF at the receiver.

Table 5.1.1 The derivation of sampling process

$m(t)$	$M(f)$
$\delta_T(t) = \delta(t - nT_s)$	$\Delta_\Omega(f) = \dfrac{1}{T}\sum\limits_{n=-\infty}^{\infty}\delta(f - nf_s)$
$m_s(t) = m(t) \cdot \delta_T(t)$	$M_s(f) = M(f) * \Delta_\Omega(f) = \dfrac{1}{T}\sum\limits_{-\infty}^{\infty}M(f - nf_s)$

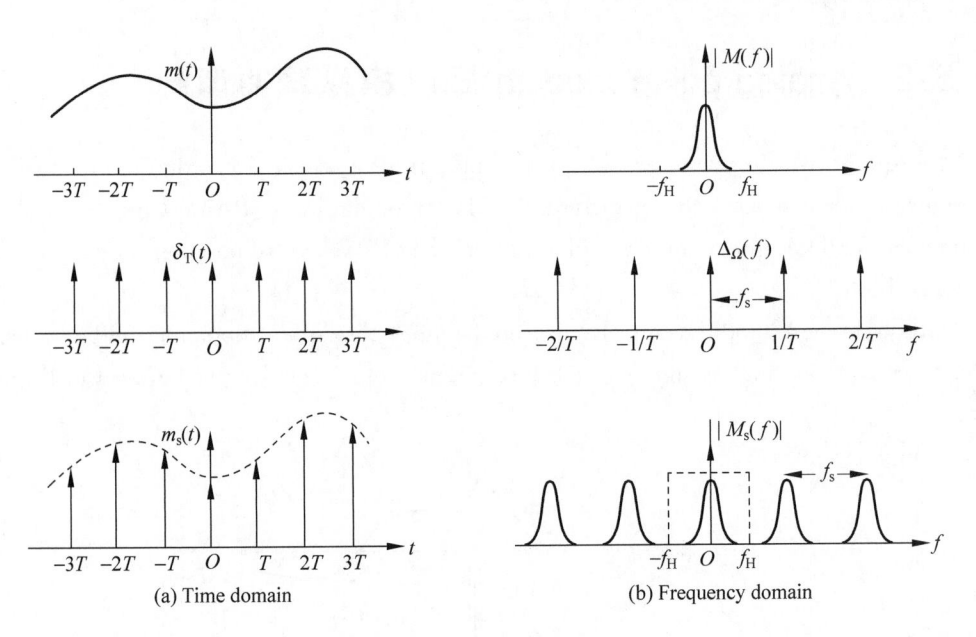

Figure 5.1.2 Sampling process

From Figure 5.1.2, the condition of restoration of original signal is $f_s \geqslant 2f_H$. The lowest sampling frequency $2f_H$ is called Nyquist sampling rate(奈奎斯特采样速率). The corresponding largest sampling time interval is called Nyquist sampling interval(奈奎斯特采样间隔).

2. The sampling of band-pass analog signals

The frequency band of band-pass signals is limited between f_L and f_H. $B = f_H - f_L$ is the bandwidth of the analog signal. Here, we only give the result. The sampling frequency f_s can be written as:

$$f_s = 2B + \frac{2KB}{n} = 2B\left(1 + \frac{K}{n}\right) \tag{5.1.1}$$

where n is the largest integer (整数) less than f_H/B, $0 < k < 1$. The relation between f_s and f_L is shown in Figure 5.1.3.

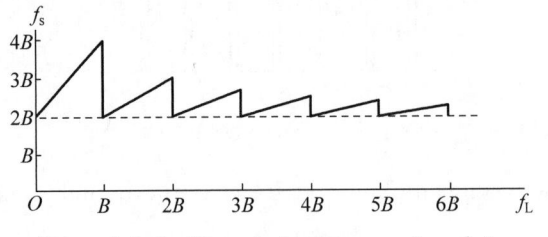

Figure 5.1.3 The relation between f_s and f_L

When $f_L = 0, f_H = 2B$, this is the sampling condition for the low pass filter.
When $f_L \gg 0, f_s \rightarrow 2B$, the signal is a narrow-band signal.

5.2 Analog pulse modulation（模拟脉冲调制）

There are three parameters of pulse: amplitude, width (duration) and position. Therefore, we can get three different pulse modulations: PAM (pulse amplitude modulation), PDM (pulse duration modulation) and PPM (pulse position modulation), as shown in Figure 5.2.1.

Although these modulations are discrete in time, they are still analog modulations. In order to convert an analog signal to a digital signal, quantizer should be used in the next step.

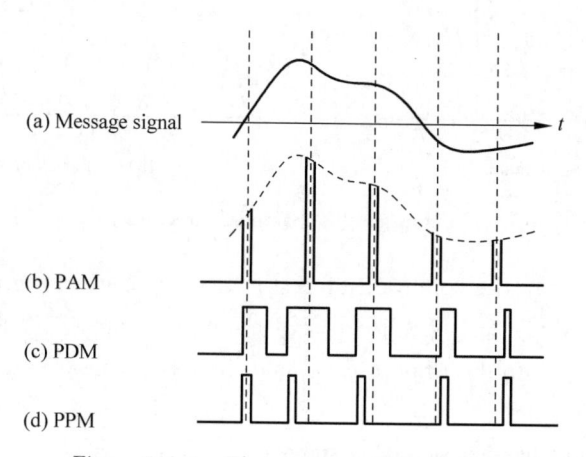

(a) Message signal

(b) PAM

(c) PDM

(d) PPM

Figure 5.2.1　Three types pulse modulation

In this part, the PAM is introduced in detail. There are two types PAM: natural sampling（自然采样）and flat-top sampling（平顶采样）.

1. Natural sampling PAM

As can be seen from Figure 5.2.2, the top of each pulse is natural.

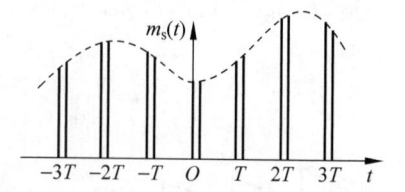

Figure 5.2.2　Natural sampling PAM

The differences between sampling and PAM is the period signal $s(t)$, which is the period pulses.

According to

$$m_s(t) = m(t) \cdot s(t)$$

Its corresponding Fourier transform is

$$M_s(f) = M(f) * S(f)$$

$$= \frac{A\tau}{T} \sum_{n=-\infty}^{\infty} \mathrm{sinc}(\pi n\tau f_{\mathrm{H}}) M(f - 2f_{\mathrm{H}})$$

Because the carrier wave is periodic rectangular pulses, the envelope of $M_s(f)$ is the sinc function. The original signal $M(f)$ can be recovered/reconstructed when $M_s(f)$ passes through a LPF at the receiver (see Figure 5.2.3).

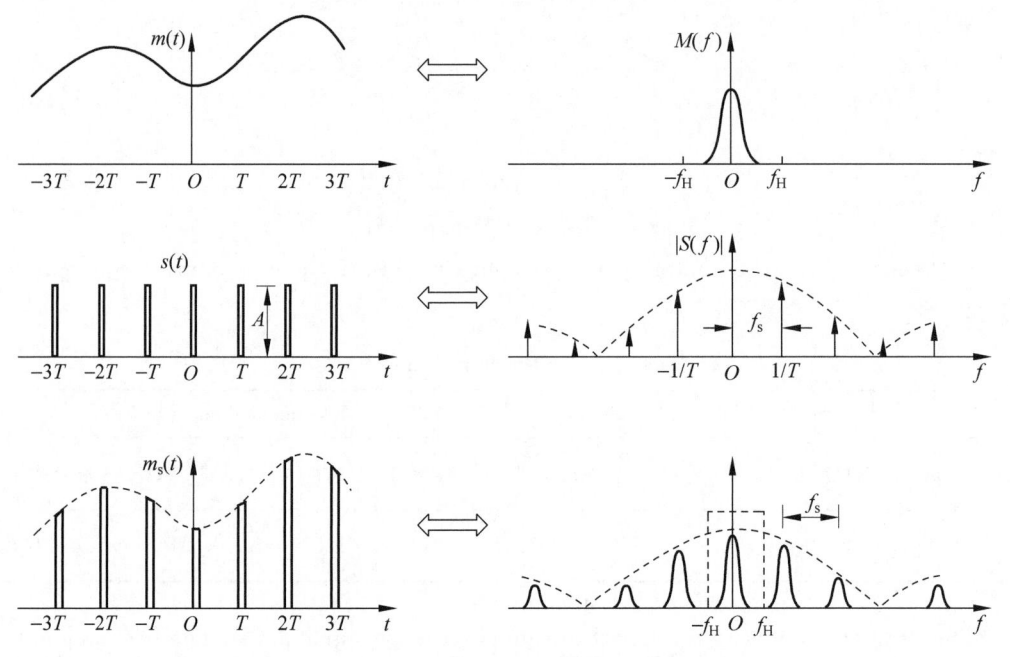

Figure 5.2.3　The waveform and spectrum of PAM

2. Flat-top sampling PAM

The top of PAM is flat, as shown in Figure 5.2.4.

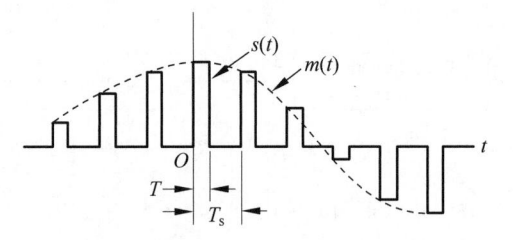

Figure 5.2.4　Flat-top sampling PAM

There are two operations involved in the generation of the PAM signal, as shown in Figure 5.2.5.

(1) Instantaneous sampling (瞬时采样).

The sampling rate f_s is choosed in accordance with the sampling theorem.

(2) Lengthening (保持) the duration of each sample so obtained to some constant value T.

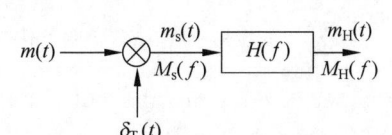

Figure 5.2.5 The principle of PAM

Where

$$h(t) = \begin{cases} 1, & 0 < t < T \\ \dfrac{1}{2}, & t = 0, t = T \\ 0, & \text{otherwise} \end{cases}$$

$$H(f) = T\operatorname{sinc}(fT)\exp(-\mathrm{j}\pi fT)$$

In Table 5.2.1, the comparisons between time and frequency domain equations are given.

Table 5.2.1 The derivation of PAM

Time domain	Frequency domain
$m_s(t) = m(t) \cdot \delta_T(t) = \displaystyle\sum_{n=-\infty}^{\infty} m(nT_s) \cdot \delta(t - nT_s)$	$M_s(f) = M(f) * \Delta_n(f) = f_s \displaystyle\sum_{n=-\infty}^{\infty} M(f - nf_s)$
$m_H(t) = m_s(t) * h(t)$	$M_H(f) = M_s(f) \cdot H(f) = f_s \displaystyle\sum_{n=-\infty}^{\infty} M(f - nf_s) \cdot H(f)$

The problem is how to recover the original message signal $m(t)$. The first step is that $m_H(t)$ passes through a LPF(low pass filter). The spectrum of LPF output is equal to $M(f)H(f)$. Obviously, there is frequency distortion because of $H(f)$. This distortion may be corrected by connecting an equalizer in cascade（串联）with the LPF, as shown in Figure 5.2.6.

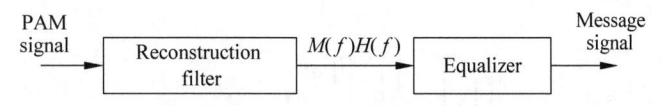

Figure 5.2.6 The reconstruction of message signal

The magnitude response of the equalizer is given by

$$\frac{1}{|H(f)|} = \frac{1}{T\operatorname{sinc}(fT)} = \frac{\pi f}{\sin(\pi fT)}$$

For a duty cycle（占空比）$T/T_s \leqslant 0.1$, the amplitude distortion is less than 0.5%, in which case the need for equalization may be omitted altogether.

5.3 Quantization process（量化过程）of sampled signal

视频讲解

Amplitude quantization：the process of transforming the sample amplitude $m(kT)$ of a message signal $m(t)$ at time $t = kT$ into a discrete amplitude $m_q(kT)$ taken from a finite set of possible amplitudes(有限个可能的取值). An example of quantization process is shown in Figure 5.3.1.

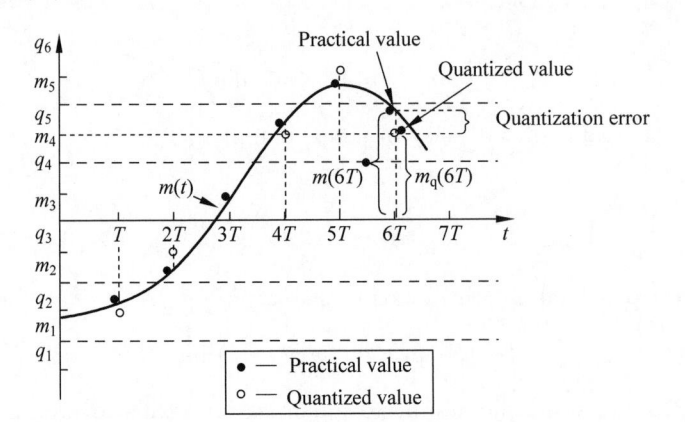

Figure 5.3.1 The quantization process

$m(kT)$ expresses the sampled signal input to a quantizer, $m_q(kT)$ expresses the quantized value of the output signal of this quantizer, as shown in Figure 5.3.2.

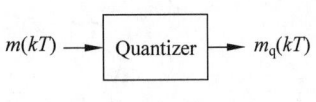

Figure 5.3.2 Quantizer

$q_1 \sim q_6$ are 6 possible levels of the quantized signal.

$m_1 \sim m_5$ are the boundary points of the quantization intervals.

$$m_q(kT) = q_i, \quad m_{i-1} \leqslant m(kT) \leqslant m_i \tag{5.3.1}$$

In this example, M intervals are with equal space, so it is called uniform quantization (均匀量化). If the M intervals are with unequal spaces, it is called nonuniform quantization (非均匀量化), which is based on the uniform quantization.

nonuniform quantization ⇒ uniform quantization

5.3.1 Uniform quantization (均匀量化)

Assume the magnitudes of the sampled signal is within $a \sim b$, and the number of quantization levels is M, so the quantization interval (量化间隔) is $\Delta \nu = \dfrac{b-a}{M}$ (also called step-size: 步长), and the boundary points of the quantization intervals are $m_i = a + i \Delta \nu$, $i = 0, 1, \cdots, M$.

Obviously, quantized output m_q is different from signal sample before quantization m_k, there is an error of quantizer. This error is usually called quantization noise (量化噪声) and signal power to quantization noise power ratio (信噪比) is used for estimating the influence of the error to the signal.

The average value N_q of the quantized noise power for uniform quantization can be expressed by

$$N_q = [(m_k - m_q)^2] = \int_a^b (m_k - m_q)^2 f(m_k) \mathrm{d}m_k$$

$$= \sum_{i=1}^{M} \int_{m_{i-1}}^{m_i} (m_k - q_i)^2 f(m_k) \mathrm{d}m_k \qquad (5.3.2)$$

where $f(m_k)$ is the probability density of signal sample m_k

$$m_i = a + i\Delta\nu$$

$$q_i = a + i\Delta\nu - \frac{\Delta\nu}{2} \qquad (5.3.3)$$

The average power of signal m_k can be expressed as

$$S = E(m_k^2) = \int_a^b m_k^2 f(m_k) \mathrm{d}m_k \qquad (5.3.4)$$

Example 5.3.1： Assume the number of quantization levels for a uniform quantizer is M, and the sample of the input signal has uniform probability density in the interval $[-a, a]$. Find the average signal to quantization noise ratio for the quantizer.

Solution：From equation(5.3.2), we can get

$$N_q = \sum_{i=1}^{M} \int_{m_{i-1}}^{m_i} (m_k - q_i)^2 f(m_k) \mathrm{d}m_k = \sum_{i=1}^{M} \int_{m_{i-1}}^{m_i} (m_k - q_i)^2 \left(\frac{1}{2a}\right) \mathrm{d}m_k$$

$$= \sum_{i=1}^{M} \int_{-a+(i-1)\Delta\nu}^{-a+i\Delta\nu} \left(m_k + a - i\Delta\nu + \frac{\Delta\nu}{2}\right)^2 \cdot \left(\frac{1}{2a}\right) \mathrm{d}m_k$$

$$= \sum_{i=1}^{M} \left(\frac{1}{2a}\right) \cdot \left(\frac{\Delta\nu^3}{12}\right) = \frac{M(\Delta\nu)^3}{24a}$$

According to $M \cdot \Delta\nu = 2a$, so the quantization noise power is

$$N_q = \frac{(\Delta\nu)^2}{12}$$

The average power of signal is

$$S_0 = \int_{-a}^{a} m_k^2 \left(\frac{1}{2a}\right) \mathrm{d}m_k = \frac{M^2}{12} \cdot (\Delta\nu)^2$$

The the average signal to quantization noise ratio is：

$$\frac{S_0}{N_q} = M^2 \quad \text{or} \quad \frac{S_0}{N_q}\bigg|_{\mathrm{dB}} = 20\lg M (\mathrm{dB})$$

$\dfrac{S_0}{N_q}$ is increasing along with the increasing of number M. That is to say when the signal is small, signal to quantization noise ratio is also small. For the speech signal, there are lots of small components. To overcome this disadvantage, the nonuniform quantization is often used in practical applications.

5.3.2　Nonuniform quantization

A nonuniform quantizer is equivalent to passing the sampled signal through a compressor and then applying the compressed signal to a uniform quantizer, as illustrated in Figure 5.3.3.

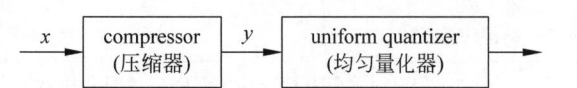

Figure 5.3.3   The equivalent process of nonuniform quantizer

Compression uses a nonlinear circuit to convert input voltage x to output voltage y:

$$y = f(x)$$

where x has nonuniform scale and y has uniform scale.

Theoretically, $f(x)$ should be a logarithm(对数)function. $f(x)$ needs to be properly modified in different requirements of the practical condition. There are two ITU recommended logarithm compression laws (对数压缩定律): A-law and μ-law, for telephone signal. The logarithm laws are give in Table 5.3.1.

Table 5.3.1   Logarithm compression laws

A-law	μ-law
$y = \begin{cases} \dfrac{Ax}{1+\ln A}, & 0 < x \leqslant \dfrac{1}{A} \\[2mm] \dfrac{1+\ln Ax}{1+\ln A}, & \dfrac{1}{A} \leqslant x \leqslant 1 \end{cases}$	$y = \dfrac{\ln(1+\mu x)}{\ln(1+\mu)}$
13 segments	15 segments
E1: China & European countries	**T1**: North American, Japan and Korea

A-law 13 segments method: A-law compression is a continuous smooth curve. It is difficult to be accurately achieved by electronic circuit, while the 13 segments method can be easily achieved by digital circuit approximately, which is plotted in Figure 5.3.4.

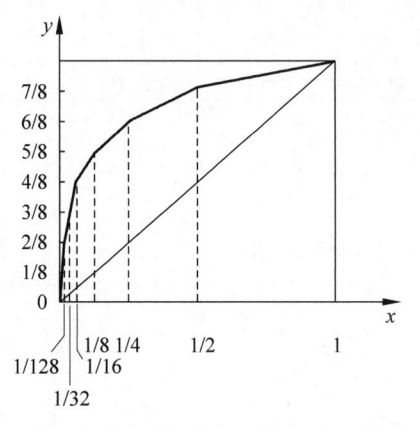

Figure 5.3.4   A-law 13 segments

In Table 5.3.2, the slope of each segment is given. As can be seen, the first and second segments have the same slope.

Table 5.3.2   Slope of each segment

No. (折线段号)	1	2	3	4	5	6	7	8
slope(斜率)	16	16	8	4	2	1	1/2	1/4

Why we called it 13 segments？

The input voltage of speech signal has positive and negative polarities. So the above compression curve is only one half of the practical characteristic curve of the compressor. There is another half in the 3rd quadrant（象限），so the curve is odd symmetry to the origin（关于原点奇对称）.

The slopes of the 1st and 2nd segments in 1st quadrant and 3rd quadrant are the same. These 4 segments compose a straight line. Therefore，it is called 13 segments.

Signal after quantization is a digital signal with discrete values. The next step is how to encode these discrete values.

5.4 PCM（pulse code modulation，脉冲编码调制）

PCM is the most basic form of digital pulse modulation.

PCM：a message signal is represented by a sequence of coded pulses which is accomplished by representing the signal in discrete form in both time and amplitude（PCM 编码后的信号在时间和幅度上的取值都是离散的）.

The most often used code is to use the binary symbols representing discrete values，such as 0 and 1. The basic operations performed in the transmitter of a PCM system are sampling，quantizing and encoding，as illustrated in Figure 5.4.1. The quantizing and encoding operations are usually performed in the same circuit，which is called an analog-to-digital converter（A/D 转换器）. The basic operations in the receiver is the regeneration of the message signals：decoding and reconstruction.

Figure 5.4.1　The PCM process

5.4.1 The principle of PCM

There are 8 bits in PCM code word（码字），which can satisfy the speech communication quality.

c_1—the polarity bit（极性位）$\begin{cases} +，c_1=1 \\ -，c_1=0 \end{cases}$；

$c_2c_3c_4$—segment bits（段落码）：000～111；

$c_5c_6c_7c_8$—inner-segment bits（段内码）：000～1111（16 inner-segments in each segment）.

Inner-segment bits are uniformly encoded according to quantization intervals，but quantization intervals of different segments are different.

Among these 8 segments，the lengths of 1st and 2nd segments are the shortest 1/128， and slopes of them are the largest, after it is equally divided into 16 sub-segments, the each sub-segment equals:

$$\Delta = \frac{1}{128} \times \frac{1}{16} = \frac{1}{2048}$$

which is called the smallest quantization interval(最小量化间隔).

If uniform quantization is used to keep the same dynamic range $1/2048$, then 11 bits code word is necessary.

While for the PCM systems, only 7 bits (不含极性位) are used, which embody the advantage of nonuniform quantization.

The typical sampling rate of telephone signal is 8kHz ($\geqslant 2 \times 3.4$kHz). Thus the typical transmission bit rate of digital telephone is 64kb/s in PCM system. This rate has been adopted by ITU-T recommendation.

Example 5.4.1: Using A-law 13 segments encoding, if the sampling value is $+1270$, find the output of PCM encoder and the quantization error.

Answer: (1) First, since $+1270 > 0$, the polarity bit $c_1 = 1$;

(2) Second, 1270 is in the 8th segment, so $c_2 c_3 c_4 = 111$;

(3) Finally, determine which inner-segment 1270 is in. There are usually two methods. One is the successive comparison (逐次比较). Another one is direct calculation. The 8th segment is plotted in Figure 5.4.2.

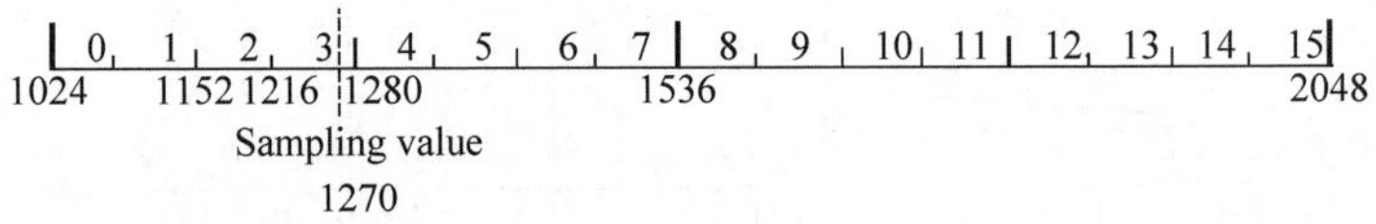

Figure 5.4.2 The quantization interval of 8th segment

Successive comparisons:

$$1270 < 1536, \quad c_5 = 0$$
$$1270 < 1280, \quad c_6 = 0$$
$$1270 > 1152, \quad c_7 = 1$$
$$1270 > 1216, \quad c_8 = 1$$

Direct calculation:

inner-segment quantization interval is $(2048 - 1024)/16 = 64$;

Since $(1270 - 1024)/64 \approx 3.8$, the inner segment coding is $c_5 c_6 c_7 c_8 = 0011$.

Quantization error is: $1270 - (1216 + 1280)/2 = 22\Delta$.

Therefore, the output of PCM encoder is 11110011 and the quantization error is 22Δ.

5.4.2 Noise in PCM system

The derivation signal to quantization noise ratio of the uniform quantizer has been discussed in Section 5.3.1.

$$S/N_q = M^2$$

When N bits binary code word is used for encoding, the above equation can be written as

$$S/N_q = 2^{2N} \quad (M = 2^N)$$

This equation shows that S/N_q of PCM system is only related to N and increases with N exponentially. For a low-pass signal, the sampling rate should be no less than $2f_H$, so in the PCM system, this is equivalent to not less than $2Nf_H$ b/s. The system bandwidth is at least $B = Nf_H$.

$$S/N_q = 2^{2(B/f_H)}$$

The S/N_q of PCM system increases with the bandwidth B exponentially.

5.4.3 Delta modulation（增量调制）

1. Principle

Delta modulation(DM or ΔM)provides a staircase approximation to the over-sampled (i. e., at a rate much higher than the Nyquist rate) version of the message signal, as illustrated in Figure 5.4.3.

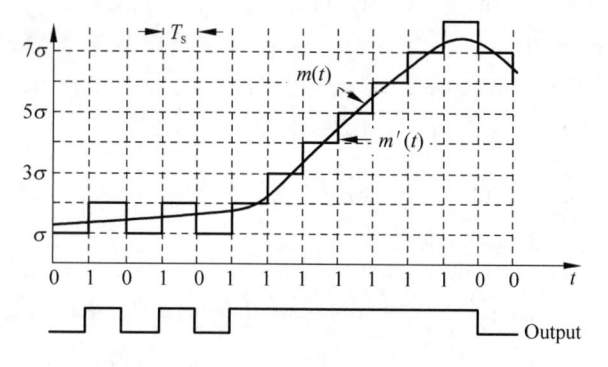

Figure 5.4.3   The waveform of DM

The difference between the input and the approximation is quantized into only two levels, namely, $\pm\Delta$, corresponding to positive and negative differences.

If the approximation falls below the signal at any sampling epoch, it is increased by Δ. If the approximation lies above the signal, it is diminished by Δ.

The principal（主要的）virtue of delta modulation is its simplicity. It may be generated by applying the sampled version of the incoming message signal to a modulator that involves a comparator（比较器）, quantizer and accumulator interconnected as shown in Figure 5.4.4. The block labeled z^{-1} inside the accumulator represents a unit delay, that is, a delay equal to one sampling period.

The basic principle of Δ modulation is in the following:

$$e[n] = m[n] - m_q[n-1]$$
$$e_q = \Delta \operatorname{sgn}[e(n)]$$
$$m_q[n] = m_q[n-1] + e_q[n]$$

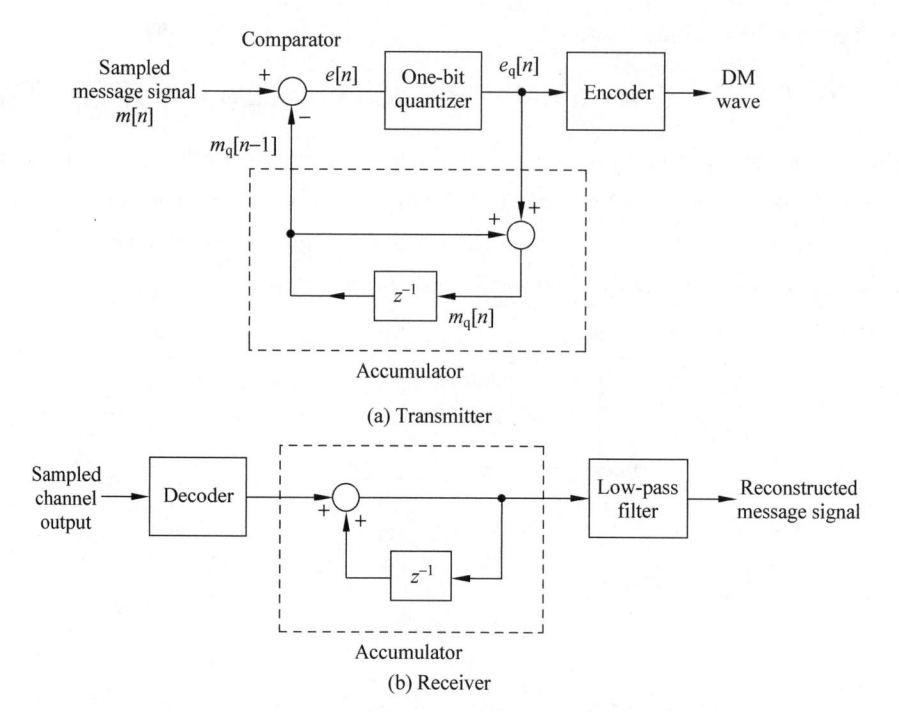

(a) Transmitter

(b) Receiver

Figure 5.4.4   DM system

The quantizer output $m_q[n]$ is coded to produce the DM signal. The rate of information transmission is simply equal to the sampling rate $f_s = \dfrac{1}{T_s}$.

2. Quantization error

There are two types of quantization error: slope overload distortion and granular noise (see Figure 5.4.5).

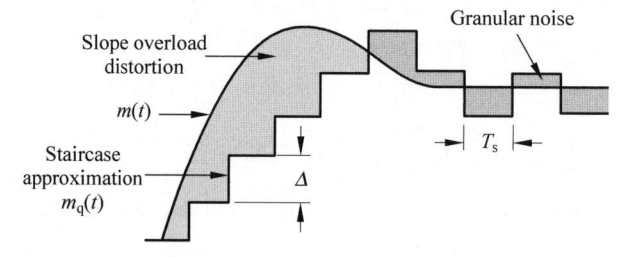

Figure 5.4.5   Two different forms of quantization error

If considered the maximum slope of the original input waveform $m(t)$, it is clear that in order for the sequence of samples $\{m_q[n]\}$ to increase as fast as the input sequence of samples $\{m[n]\}$ in a region of maximum slope of $m(t)$, we require that the following condition:

$$\frac{\Delta}{T_s} \geqslant \max \left| \frac{\mathrm{d}m(t)}{\mathrm{d}t} \right|$$

be satisfied. Otherwise, we find the step-size Δ is too small for the staircase approximation $m_q(t)$. In contrast to slope-overloaded distortion, granular noise is analogous (类似) to quantization noise in a PCM system.

5.4.4 DPCM(differential pulse code modulation，差分脉冲调制)

When a voice or video signal exhibits a high degree of correlation between adjacent samples （邻近的采样）the resulting PCM encoded signal will contain redundant information（冗余信息）.

In order to remove this redundancy before encoding，we use a more efficient coded signal，which is the basic idea of DPCM，as shown in Figure 5.4.6.

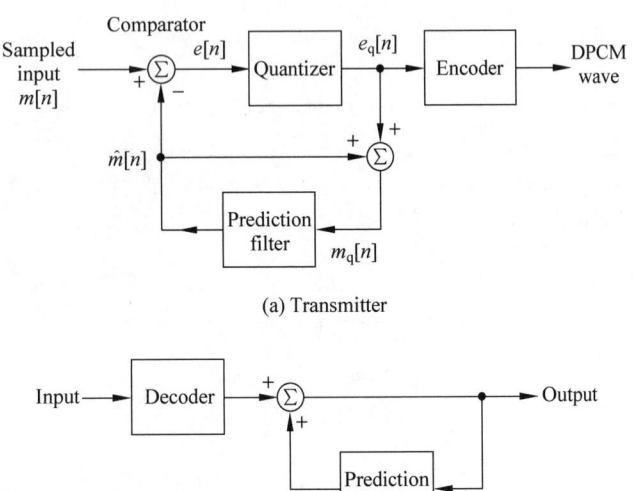

(a) Transmitter

(b) Receiver

Figure 5.4.6　DPCM system

The input signal of the DPCM quantizer is

$$e[n]=m[n]-\hat{m}[n]$$

where $m[n]$ is the unquantized sample signal and $\hat{m}[n]$ is the prediction of $m[n]$.

The relationship between the output and input of the predictor is

$$\hat{m}=\sum_{i=1}^{q}a_i m_q（线性预测）$$

where p is the prediction order and a_i is the prediction coefficient（系数）. The predicted value is the weighted sum of previous p samples of the signal with quantization error.

If $p=1,a_1=1$ then $\hat{m}=m_{q-1}$. The predictor is simply a delay circuit，and the delay time is sampling interval T. If the quantizer is one-bit（two-level），the DPCM is ΔM，i.e. ΔM is the special case of DPCM.

5.5　TDM(time-division multiplexing，时分复用)

The concept of TDM is shown in Figure 5.5.1. Each input message signal is first restricted in a low-pass anti-aliasing filter. The functions of each part are described below.

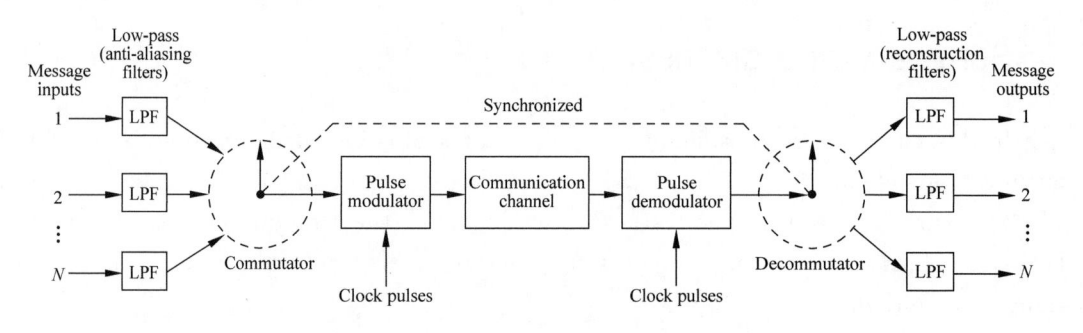

Figure 5.5.1　Block diagram of TDM system

LPF：restricts each input message in bandwidth and removes the nonessential frequencies.

Commutator（换向器）：which is usually implemented using electronic switching circuitry. The function is to sequentially interleave these N samples inside. The sampling interval is T. It is the essence of TDM operation.

Pulse modulator：transform the multiplexed signal into a form suitable for transmission over the common channel.

Pulse demodulator：the reverse operation of pulse modulator.

Decommutator：operates in synchronism（同步）with commutator in the transmitter.

The multiplexing of digital signals is accomplished by using a bit-by-bit interleaving procedure with a selector switch that sequentially takes a bit from each incoming line and then applies it to the high-speed common line. At the receiving end of the system the output of this common line is separated out into its low-speed individual components and then delivered to their respective destination，as plotted in Figure 5.5.2.

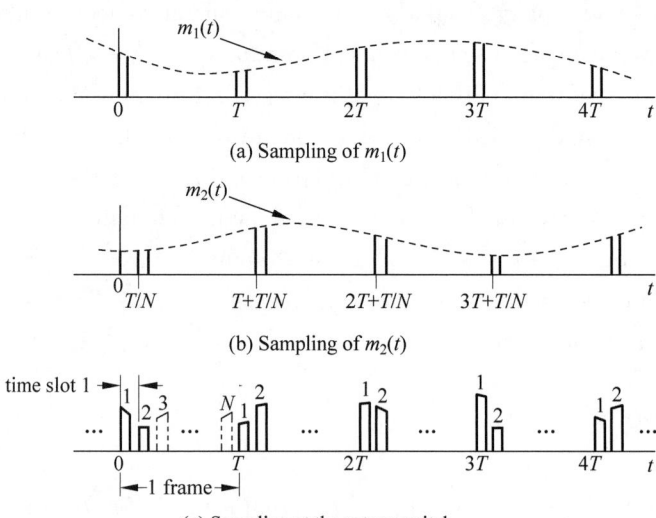

(a) Sampling of $m_1(t)$

(b) Sampling of $m_2(t)$

(c) Sampling at the rotary switch

Figure 5.5.2　Principle of TDM

Summary and discussion

In this chapter，we introduced the process of analog signal digitization：sampling，quantization and encoding.

（1）Sampling，which operates in the time domain；the sampling process is the link between an analog waveform and its discrete-time representation，which also is an analog signal in amplitude.

（2）Quantization，which operates in the amplitude domain；the quantization process is the link between an analog waveform and its discrete-amplitude representation.

（3）Encoding，which operates in both time and amplitude domain，encoding process is the link between discrete-amplitude and binary representation.

The sampling process is based on the sampling theorem，which states that when an analog signal with frequency band within（0，f_H）is sampled，the lowest sampling rate should not be less than Nyquist sampling rate $2f_H$. The sampled signal may be converted to different analog pulse modulation signals，including PAM，PDM and PPM. In TDM of several channels，signal processing usually begins with PAM. In order to make full use of the time interval of each sampling point，TDM is designed and applied in speech communication system.

There are two methods for the quantization of a sampled signal，one is uniform quantization，another one is nonuniform quantization. Nonuniform quantization is usually applied in speech signal with the logarithm characteristic recommended by ITU-T，i. e.，A-law and μ-law，which may effectively improve signal to quantization noise ratio，especially the small amplitude signal. European countries and China adopt A-law；North American countries and Japan as well as other counties and areas adopt μ-law. 13 segment and 15 segment methods are applied in digital circuits to achieve the A-law and μ-law quantization.

Signal after quantization is already a digital signal. Encoding methods，such as PCM，DPCM and ΔM，are usually used to convert a quantized signal into a binary signal. This process is lossy in the sense that some information is lost，but the loss of information is under the designer's control in that it can be made small enough.

The signal to quantization ratio of PCM system increases with the bandwidth B exponentially，while the analog modulation increases linearly. That is why PCM and its improved coding are widely used.

Homework

5.1　Suppose the spectrum of a message signal $m(t)$ is $M(f)$，its expression is

$$M(f) = \begin{cases} 1 - \dfrac{|f|}{200}, & |f| < 200\mathrm{Hz} \\ 0, & \text{otherwise} \end{cases}$$

(1) If the sampling rate is 300Hz, try to draw the spectrum of the sampled signal $m_s(t)$;

(2) If the sampling rate is 400Hz, try to re-draw it.

5.2   Using A-law 13 segment encoding, if the sampling value is 635, find the output of PCM encoder.

5.3   Suppose the message signal is $m(t) = 9 + A\cos\omega t$, where $A \leqslant 10$V. If $m(t)$ is quantized to 40 levels, try to find the number of binary bits and the quantization intervals.

Terminologies

sampling	采样	commutator	换向器
quantizing	量化	PCM	脉冲编码调制
encoding	编码	DPCM	差分脉冲编码调制
uniform quantization	均匀量化	ΔM	增量调制
nonuniform quantization	非均匀量化	PAM	脉冲幅度调制
quantization noise	量化噪声	PDM	脉冲宽度调制
equalizer	均衡器	PPM	脉冲位置调制
compressor	压缩器	TDM	时分复用

Baseband pulse transmission

Mind map:

Mind map for

Chapter 6

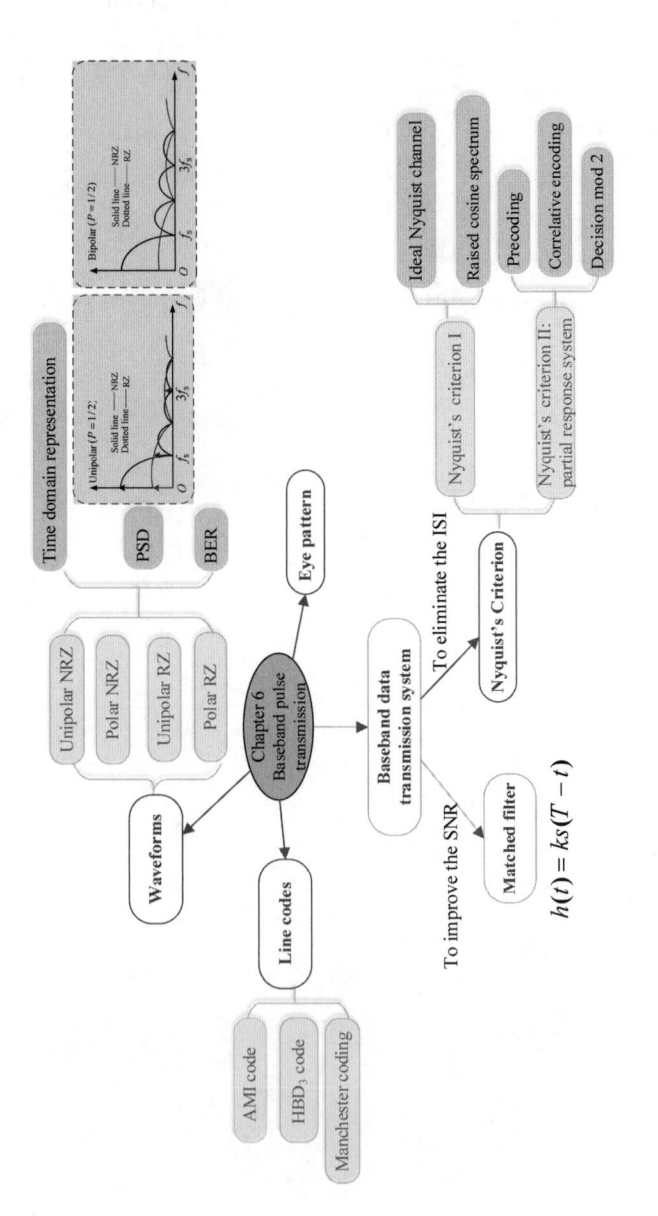

In this chapter，we study the transmission of digital data over a baseband channel，with emphasis on the following content：

- Waveform and representation of baseband digital signals.
- Symbol code types of baseband digital signal，mainly discuss AMI code and HDB₃ code.
- Matched filter（匹配滤波器），which is the optimum system for detecting a known signal in additive white Gaussian noise.
- Discuss the bit error rate（误码率）of unipolar and polar baseband systems.
- Intersymbol interference（ISI，码间串扰），which arises when the channel is dispersive as is commonly the case in practice.
- Nyquist's criterion(奈奎斯特准则)．

  Nyquist's criterion Ⅰ：the transfer characteristic of no intersymbol interference.

  Nyquist's criterion Ⅱ：partial response system（部分响应系统）．
- Eye pattern(眼图)．

6.1　Waveform and frequency characteristics of baseband digital signal

视频讲解

Binary signals are denoted by 0 and 1. Rectangular voltage pulse is usually used as the waveform. The waveform of baseband digital signal is also called unmodulated digital signal.

6.1.1　Waveform of baseband digital signal

Waveform can be used for the electrical representation of a binary data stream. Figure 6.1.1 displays the waveforms of five important line codes for the example data stream 0 1 0 1 1 0 0 0 1.

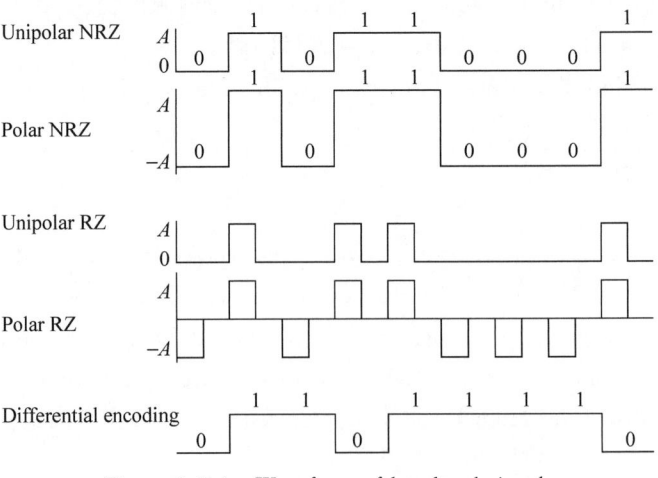

Figure 6.1.1　Waveform of baseband signals

1. Unipolar nonreturn-to-zero（Unipolar NRZ：单极性不归零）

Symbol 1 is represented by transmitting a pulse of amplitude A for the duration of the symbol，and 0 is represented by switching off the pulse. This code is also referred to as on-off signaling.

2. Polar nonreturn-to-zero（Polar NRZ：双极性不归零）

Symbol 1 and 0 are represented by transmitting pulses of amplitudes $+A$ and $-A$, respectively.

3. Unipolar return-to-zero（Unipolar RZ：单极性归零）

Symbol 1 is represented by a rectangular pulse of amplitude A and half-symbol width, and symbol 0 is represented by transmitting no pulse.

4. Polar return-to-zero（Polar RZ：双极性归零）

Both 1 and 0 have half-symbol width. One is positive and another is negative.

5. Differential encoding（差分编码）

A transition is used to designate symbol 1 in the coming binary data stream，while no transition is used to designate symbol 0.

As can be seen from Figure 6.1.1, when 1 appears, the voltage jumps; when 0 appears，the voltage doesn't change. Of course, the rule may be reversed.

6.1.2　Symbol code types of baseband digital signals for transmission

The waveform of original digital signal has been discussed. In order to adapt to transmission in the channel，the symbol code types of these waveforms should be converted. Some requirements on the symbol code types are as follows：

- No D.C. component（无直流分量）and few low frequency components；
- Containing timing information（含有定时信息）；
- Certain error-detecting ability；
- Suitable for various information sources（能适应信源的变化）；
- Simplicity of encoding and decoding（编译码简单）.

Here，three code types for transmission are introduced.

1. AMI code

Message symbol：	0	1	0	1	1	0	0	0	1
AMI code：	0	+1	0	-1	+1	0	0	0	-1

Encoding rule：1 in message symbol will be converted into $+1$ and -1 alternatively；0 will keep 0. Because $+1$ and -1 appear alternatively，there is no D.C. component. Its decoding circuit is very simple. Disadvantage：when a long string of 0 appears, it is difficult to get timing information. HDB_3 code can overcome this disadvantage.

2. HDB_3（the 3rd order High Density Bipolar code：三阶高密度双极性码）

Message symbol	1	0	0	0	0	1	0	0	0	0	1	1	0	0	0	0	1
AMI code	+1	0	0	0	0	-1	0	0	0	0	+1	-1	0	0	0	0	+1
HDB_3 code	+1	0	0	0	+V	-1	0	0	0	-V	+1	-1	+B	0	0	+V	-1

Coding rule：

（1）When there are no more than 4（not including 4）continuous 0s，no change will be made，AMI is HDB$_3$ code.

（2）When there are more than 4 continuous 0s，the fourth 0 will be converted into a symbol which has the same polarity of the previous nonzero symbol. This symbol is called violation symbol（破坏码）. The requirement of adjacent V_S is "polarity alternatively inverse（极性交替）".

（3）If（2）can not be satisfied，use "regulating impulse(调节脉冲)" B，the sign of B is the inverse of the previous nonzero symbol.

The decoding of HDB$_3$ is simple：V and its previous 3 continuous symbols must be continuous 0s，then 4 continuous 0s may be restored.

3. Manchester code（曼彻斯特码）

Its coding rule is that each binary symbol is converted into a period of square wave with different phases. Advantages of this code is：no D. C. component；a lot of timing information；simple coding method. The disadvantage is that it occupied double frequency band.

Message symbol　　　　1　1　0　0　1　0　1

Manchester code　　　10　10　01　01　10　01　10

6.1.3　Frequency characteristic

Assume the signal pulse is rectangular and a sign single pulse is $g(t)$，the frequency characteristic（PSD）of a random binary signal sequence will be discussed.

Symbols 0 and 1 can be represented by $g_1(t)$ and $g_2(t)$，with occurring probabilities P and $1-P$，then signal sequence $s(t)$ can be expressed by

$$s(t) = \sum_{n=-\infty}^{\infty} s_n(t) \tag{6.1.1}$$

where

$$s_n(t) = \begin{cases} g_1(t-nT) & \text{probability} \quad P \\ g_2(t-nT) & \text{probability} \quad (1-P) \end{cases}$$

$s(t)$ can be regarded as the combination of a stationary wave $v_c(t)$（稳态波）and an alternating wave $u_c(t)$（交变波）.

$$s(t) = v_c(t) + u_c(t) \tag{6.1.2}$$

The PSD of $v(t)$ is

$$P_v(f) = \sum_{m=-\infty}^{\infty} |f_c[PG_1(mf_c) + (1-P)G_2(mf_c)]|^2 \delta(f-mf_c) \quad \text{— 离散谱} \tag{6.1.3}$$

where $G_1(mf_c) = \int_{-\infty}^{\infty} g_1(t)e^{-j2\pi mf_c t} \mathrm{d}t$；$G_2(mf_c) = \int_{-\infty}^{\infty} g_2(t)e^{-j2\pi mf_c t} \mathrm{d}t$，$f_c = \dfrac{1}{T}$，$T$ is the symbol width（码元宽度）.

The PSD of $u(t)$ is

$$P_u(f) = f_c p(1-p) |G_1(f) - G_2(f)|^2 \quad \text{——连续谱} \tag{6.1.4}$$

So the PSD of $s(t)$ is

$$P_s(f) = P_v(f) + P_u(f) \tag{6.1.5}$$

Example 6. 1. 1：Suppose the pulse is rectangular，try to get the spectrum of unipolar （单极性） NRZ and RZ.

Solution：Suppose $g_1(t) = 0, g_2(t) = g(t)$，then substitute them into the following formula

$$P_s(f) = P_u(f) + P_v(f) = f_s P(1-P) |G_1(f) - G_2(f)|^2 +$$

$$\sum_{m=-\infty}^{\infty} |f_s [PG_1(mf_s) + (1-P)G_2(mf_s)]|^2 \delta(f - mf_s)$$

We can get the PSD of unipolar signal is

$$P_s(f) = f_s P(1-P) |G(f)|^2 + \sum_{m=-\infty}^{\infty} |f_s(1-P)G(mf_s)|^2 \delta(f - mf_s)$$

When $P = 1/2$,

$$P_s(f) = \frac{1}{4} f_s |G(f)|^2 + \frac{1}{4} f_s^2 \sum_{m=-\infty}^{\infty} |G(mf_s)|^2 \delta(f - mf_s)$$

（1）If the unipolar is NRZ, the time domain and frequency domain of $g(t)$ is

$$g(t) = \begin{cases} 1, & |t| \leqslant \dfrac{T_s}{2} \\ 0, & \text{otherwise} \end{cases} \quad \text{and} \quad G(f) = T_s \left(\frac{\sin \pi f T_s}{\pi f T_s} \right) = T_s \mathrm{Sa}(\pi f T_s)$$

So the PSD is

$$P_s(f) = \frac{1}{4} f_s T_s^2 \left(\frac{\sin \pi f T_s}{\pi f T_s} \right) + \frac{1}{4} \delta(f) = \frac{T_s}{4} \mathrm{Sa}^2(\pi f T_s) + \frac{1}{4} \delta(f)$$

（2）If the unipolar is RZ (duty cycle is 50%), the frequency domain of $g(t)$ is

$$G(f) = \frac{T_s}{2} \mathrm{Sa} \left(\frac{\pi f T_s}{2} \right)$$

The PSD is as follows：

$$P_s(f) = \frac{T_s}{16} \mathrm{Sa}^2 \left(\frac{\pi f T_s}{2} \right) + \frac{1}{16} \sum_{m=-\infty}^{\infty} \mathrm{Sa}^2 \left(\frac{m \pi}{2} \right) \delta(f - mf_s)$$

According to the above equation，the PSD can be plotted in Figure 6.1.2.

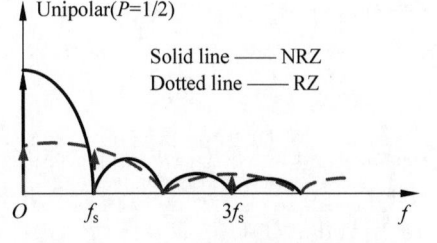

Figure 6. 1. 2   The PSD of NRZ and RZ(for unipolar)

Example 6. 1. 2: Suppose the pulse is rectangular, try to get the spectrum of polar（双极性）NRZ and RZ.

Solution: Suppose $g_1(t) = -g_2(t) = g(t)$, then according to the following formula

$$P_s(f) = P_u(f) + P_v(f) = f_s P(1-P)|G_1(f) - G_2(f)|^2 +$$

$$\sum_{m=-\infty}^{\infty} |f_s[PG_1(mf_s) + (1-P)G_2(mf_s)]|^2 \delta(f - mf_s)$$

we can get

$$P_s(f) = 4f_s P(1-P)|G(f)|^2 + \sum_{m=-\infty}^{\infty} |f_s(2P-1)G(mf_s)|^2 \delta(f - mf_s)$$

When $P = 1/2$,

$$P_s(f) = f_s|G(f)|^2$$

(1) If the polar is NRZ, $P_s(f) = T_s \text{Sa}^2(\pi f T_s)$

(2) If the polar is RZ (duty ratio is 50%), $P_s(f) = \dfrac{T_s}{4} \text{Sa}^2\left(\dfrac{\pi}{2} f T_s\right)$

The PSD can be plotted in Figure 6. 1. 3.

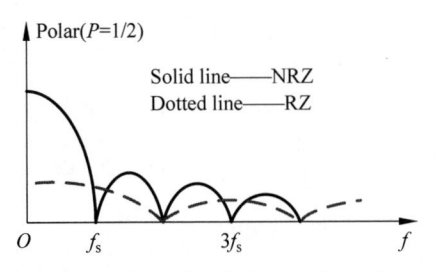

Figure 6. 1. 3　The PSD of NRZ and RZ(for polar)

6.2　Matched filter

The main index of digital communication performance is error probability. Therefore, it is appropriate to set the minimum error probability as the criterion for being "optimum".

Here we only discuss the LTI (linear time-invariant) system, as shown in Figure 6. 2. 1.

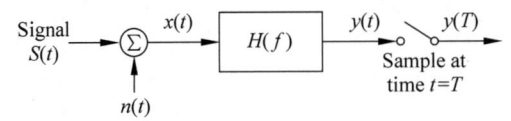

Figure 6. 2. 1　Linear receiver

The filter input $x(t)$ consists of a pulse signal $s(t)$ and additive channel noise $n(t)$, as shown by

$$x(t) = s(t) + n(t) \quad 0 \leqslant t \leqslant T \tag{6.2.1}$$

where T is an arbitrary observation interval.

The spectral density function of the incoming symbol $s(t)$ is $S(f)$.

The double-side power spectral density of the noise $n(t)$ is $P_n(f) = \dfrac{n_0}{2}$.

Since the filter is linear, the output $y(t)$ may be expressed as

$$y(t) = s_o(t) + n_o(t) \tag{6.2.2}$$

where $s_o(t) = \displaystyle\int_{-\infty}^{\infty} H(f)S(f)e^{j2\pi ft}\,\mathrm{d}f$.

The output noise power spectral density is

$$P_{n_o}(f) = H^*(f)H(f) \cdot P_{n_i}(f) = |H(f)|^2 P_{n_i}(f) \tag{6.2.3}$$

So the output noise power N_o equals to

$$N_o = \int_{-\infty}^{\infty} |H(f)|^2 \frac{n_0}{2}\mathrm{d}f = \frac{n_0}{2}\int_{-\infty}^{\infty} |H(f)|^2\mathrm{d}f \tag{6.2.4}$$

At the sampling instant t_0, the instantaneous power of the output signal to average noise power ratio is

$$r_0 = \frac{|s_o(t_0)|^2}{N_o} = \frac{\left|\displaystyle\int_{-\infty}^{\infty} H(f)S(f)e^{j2\pi ft_0}\,\mathrm{d}f\right|^2}{\dfrac{n_0}{2}\displaystyle\int_{-\infty}^{\infty} |H(f)|^2\mathrm{d}f} \tag{6.2.5}$$

To find the maximum r_0, we use the Schwarz inequality（施瓦兹不等式）：

$$\left|\int_{-\infty}^{\infty} f_1(x) \cdot f_2(x)\mathrm{d}x\right|^2 \leqslant \int_{-\infty}^{\infty} |f_1(x)|^2\mathrm{d}x \int_{-\infty}^{\infty} |f_2(x)|^2\mathrm{d}x \tag{6.2.6}$$

If $f_1(x) = kf_2^*(x)$, where k is a constant（常数）, then the equal sign in the above equation holds. Now, let $f_1(x) = H(f), f_2(x) = S(f)e^{j2\pi ft_0}$, we have

$$r_0 \leqslant \frac{\displaystyle\int_{-\infty}^{\infty} |H(f)|^2\mathrm{d}f\, |S(f)|^2\mathrm{d}f}{\dfrac{n_0}{2}\displaystyle\int_{-\infty}^{\infty} |H(f)|^2\mathrm{d}f} = \frac{\displaystyle\int_{-\infty}^{\infty} |S(f)|^2\mathrm{d}f}{\dfrac{n_0}{2}} = \frac{2E}{n_0} \tag{6.2.7}$$

when $H(f) = kS^*(f)e^{-j2\pi ft_0}$, the above equal sign holds.

The impulse response function

$$h(t) = \int_{-\infty}^{\infty} H(f)e^{j2\pi ft}\,\mathrm{d}f = ks(t_0 - t) \tag{6.2.8}$$

$h(t)$ should satisfy $h(t) = 0$, when $t < 0$. That is $s(t_0 - t) = 0$, when $t < 0$. Or satisfying $s(t) = 0$ when $t > t_0$.

Usually, $t_0 = T$. Hence, $h(t)$ can be written as：

$$\boxed{h(t) = ks(T - t)} \tag{6.2.9}$$

Example 6.2.1：Suppose the message signal is a rectangular shape(see Figure 6.2.2(a)), according to equation (6.2.9), the matched filter is plotted in Figure 6.2.2(b), and the output of the matched filter is shown in Figure 6.2.2(c). The maximum value of the output is at $t = T$.

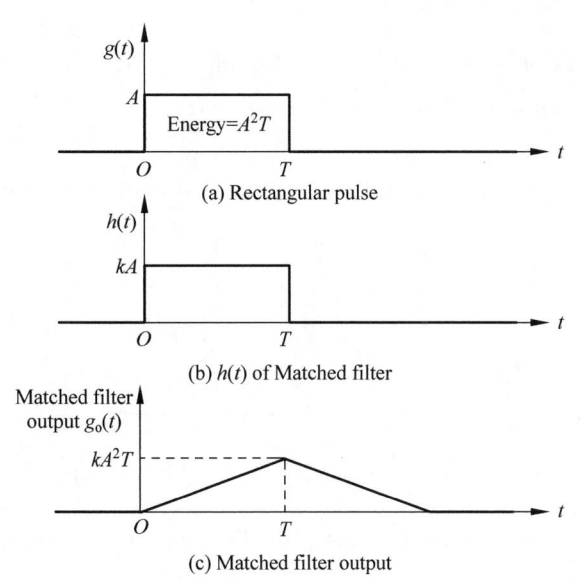

Figure 6.2.2　Matched filter

6.3　Error rate due to noise

视频讲解

The structure of receiver is shown in Figure 6.3.1.

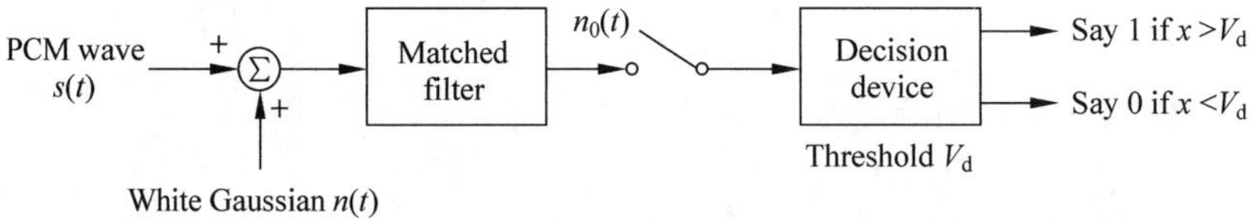

Figure 6.3.1　Receiver for baseband transmission of binary-encoded PCM wave

The input noise $n(t)$ is a stationary Gaussian random process, and matched filter is a liner receiver, therefore, the output noise $n_0(t)$ is also a stationary Gaussian random process. The probability density function of $n_0(t)$ is

$$f(v) = \frac{1}{\sqrt{2\pi}\,\delta_n} \exp\left[\frac{-v^2}{2\delta_n^2}\right] \qquad (6.3.1)$$

where mean is 0 and variance is δ_n^2.

6.3.1　Binary polar baseband system

In the signaling interval $0 \leqslant t \leqslant T$, the received signal can be written as

$$x(t) = \begin{cases} +A + n(t), & 1 \quad \text{was} \quad \text{sent} \\ -A + n(t), & 0 \quad \text{was} \quad \text{sent} \end{cases} \qquad (6.3.2)$$

After the matched filter, there is a decision making device. The sample value $x(T_s)$ should be compared to a threshold V_d.

The following question is how to get a proper V_d to get the minimum error probability.

There are two possible kinds of error to be considered, as plotted in Figure 6.3.2:

(1) Symbol 1 is chosen when a 0 was actually transmitted, this error is the error of first kind $P(1/0)$.

(2) Symbol 0 is chosen when a 1 was actually transmitted, this is the second kind error $P(0/1)$.

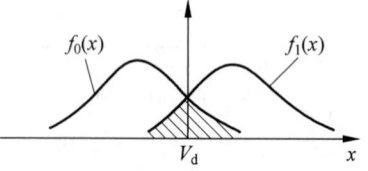

Figure 6.3.2　Noise analysis of PCM for polar signal

When symbol 0 was transmitted, the probability density function of $x(kT_s)$ is

$$f_0(x) = \frac{1}{\sqrt{2\pi}\delta_n} \exp\left[-\frac{(x+A)^2}{2\delta_n^2}\right] \qquad (6.3.3)$$

when symbol 1 was transmitted, the probability density function of $x(kT_s)$ is

$$f_1(x) = \frac{1}{\sqrt{2\pi}\delta_n} \exp\left[-\frac{(x-A)^2}{2\delta_n^2}\right] \qquad (6.3.4)$$

$$P(1/0) = P(x > V_d) = \int_{V_d}^{\infty} f_0(x)\,\mathrm{d}x = \frac{1}{2} - \frac{1}{2}\mathrm{erf}\left(\frac{V_d + A}{\sqrt{2}\delta_n}\right) \qquad (6.3.5)$$

$$P(0/1) = P(x < V_d) = \int_{-\infty}^{V_d} f_1(x)\,\mathrm{d}x = \frac{1}{2} + \frac{1}{2}\mathrm{erf}\left(\frac{V_d - A}{\sqrt{2}\delta_n}\right) \qquad (6.3.6)$$

The total P_e of the system is

$$P_e = P(1)P(0/1) + P(0)P(1/0) \qquad (6.3.7)$$

From Figure 6.3.2, when $P(1) = P(0) = 1/2$, we can find that when $V_d = 0$, P_e is the minimum.

The optimum threshold is

$$V_d^* = 0$$

Corresponding symbol error probability is

$$P_e = \frac{1}{2}\mathrm{erfc}\left(\frac{A}{\sqrt{2}\delta_n}\right)$$

6.3.2　The unipolar baseband system

The unipolar noise analysis is similar to the polar system. The different is the position of the $f_0(x)$, as shown in Figure 6.3.3.

The optimum threshold is

$$V_d^* = \frac{A}{2} + \frac{\sigma_n^2}{A}\ln\frac{P(0)}{P(1)} \qquad (6.3.8)$$

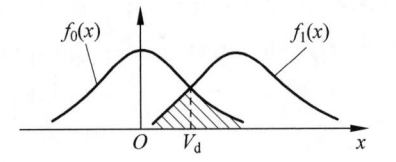

Figure 6.3.3 Noise analysis of PCM for unipolar signal

When $P(0)=P(1)=1/2$, $V_d^*=A/2$, the corresponding error probability is

$$P_e = \frac{1}{2}\mathrm{erfc}\left(\frac{A}{2\sqrt{2}\,\delta_n}\right) \tag{6.3.9}$$

Question: Which system is better? Unipolar or polar? Why?

6.4 Intersymbol Interference

Consider a baseband binary PAM system, a general form is illustrated in Figure 6.4.1.

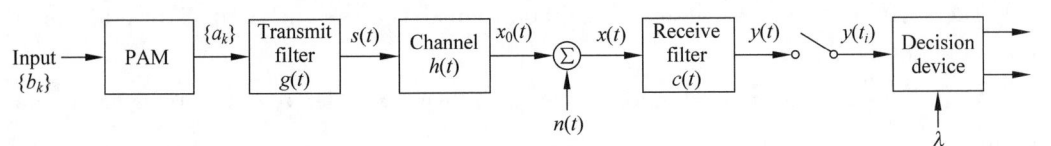

Figure 6.4.1 Baseband binary data transmission system

Intersymbol interference arises when the communication channel is dispersive (弥散).

The PAM modifies the binary sequence $\{b_k\}$ into a new sequence $\{a_k\}$, whose amplitude a_k is represented in the polar form

$$a_k = \begin{cases} +1, & \text{if symbol } b_k \text{ is } 1 \\ -1, & \text{if symbol } b_k \text{ is } 0 \end{cases} \tag{6.4.1}$$

The transmitted signal is

$$s(t) = \sum_k a_k g(t - kT_b) \tag{6.4.2}$$

where T_b is the pulse duration.

The receiver filter output can be written as:

$$y(t) = \mu \sum_k a_k p(t - kT_b) + n(t) \tag{6.4.3}$$

The scale $\mu p(t)$ is obtained by a double convolution

$$\mu p(t) = g(t) * h(t) * c(t) \tag{6.4.4}$$

Assume $p(0)=1$, the Fourier transform of $\mu p(t)$ is

$$\mu P(f) = G(f)H(f)C(f) \tag{6.4.5}$$

where $P(f)$, $G(f)$, $H(f)$, $C(f)$ are the Fourier transforms of $p(t)$, $g(t)$, $h(t)$, and $c(t)$.

So the received filter output $y(t)$ is sampled at time $t_i = iT_b$, yielding

$$y(t_i) = \mu a_i + \mu \sum_{\substack{k=-\infty \\ k \neq i}}^{\infty} a_k p[(i-k)T_b] + n(t_i) \tag{6.4.6}$$

μa_i represents the ith transmitted bit, which is the useful information. The second term represents the residual effect of all other transmitted bits on i^{th} bit, which is called ISI. The last term represents noise sample at time t_i. Under ideal conditions (in the absence of both ISI and noise):

$$y(t_i) = \mu a_i \qquad (6.4.7)$$

We have discussed the matched filter, which can get the maximum signal to noise ratio. In this part, we consider how to eliminate the ISI.

视频讲解

6.4.1　Nyquist's criterion Ⅰ

In order to eliminate ISI, $\sum_{n \neq k} a_n p\left[(k-n)T_b + t_0\right] = 0$ should be satisfied.

The time-domain condition of system overall transfer characteristic $p(t)$ is

$$p(kT_b) = \begin{cases} 1, & k = 0 \\ 0, & k \neq 0 \end{cases} \qquad (6.4.8)$$

Only the sample of $p(t)$ at the sampling point equals 1, and the samples of other pulses $p(t \pm nT)$ are equal to zero.

According to the time-domain condition, we can derive the frequency-domain condition:

$$\sum_i P\left(w + \frac{2\pi i}{T_b}\right) = T_b, \qquad |w| \leqslant \frac{\pi}{T_b} \qquad (6.4.9)$$

This condition is called **Nyquist's criterion I.**

1. Ideal Nyquist channel

If $P(f)$ has ideal rectangular characteristic, Nyquist's criterion can be satisfied.

$$P(f) = \begin{cases} T_b, & |f| \leqslant 1/2T_b \\ 0, & \text{other} \end{cases} \quad \text{and} \quad p(t) = \frac{\sin \dfrac{\pi t}{T_b}}{\dfrac{\pi t}{T_b}} = \text{Sa}(\pi t/T_b) \qquad (6.4.10)$$

Figure 6.4.2(a) and 6.4.2(b) show plots of $P(f)$ and $p(t)$. In Figure 6.4.2(a), the normalized form of the frequency function $P(f)$ is plotted for positive and negative frequencies. In Figure 6.4.2(b), the signaling intervals and corresponding centered sampling instants are included.

From the above figure, when $t = \pm kT_b$, $p(t) = 0$.

$P(t)$ has its peak value at the origin(原点) and goes through zero at integer multiples of the bit duration T_s. If the received waveform is sampled at the instant time $t = 0$, $\pm T_b$, $\pm 2T_b, \cdots$, the pulses $\mu p(t - iT_b)$ will not interfere with each other.

$$\left. \begin{array}{l} \text{Nyquist bandwidth:} \ \omega = \dfrac{1}{2T_b} \ (\text{Hz}) \\[2mm] \text{Nyquist (bit) rate:} \ R_b = 2\omega = \dfrac{1}{T_b} \ (\text{band}) \end{array} \right\} \eta = R_b/\omega = 2$$

However, the ideal low-pass transmission characteristic can not be physically achieved. The

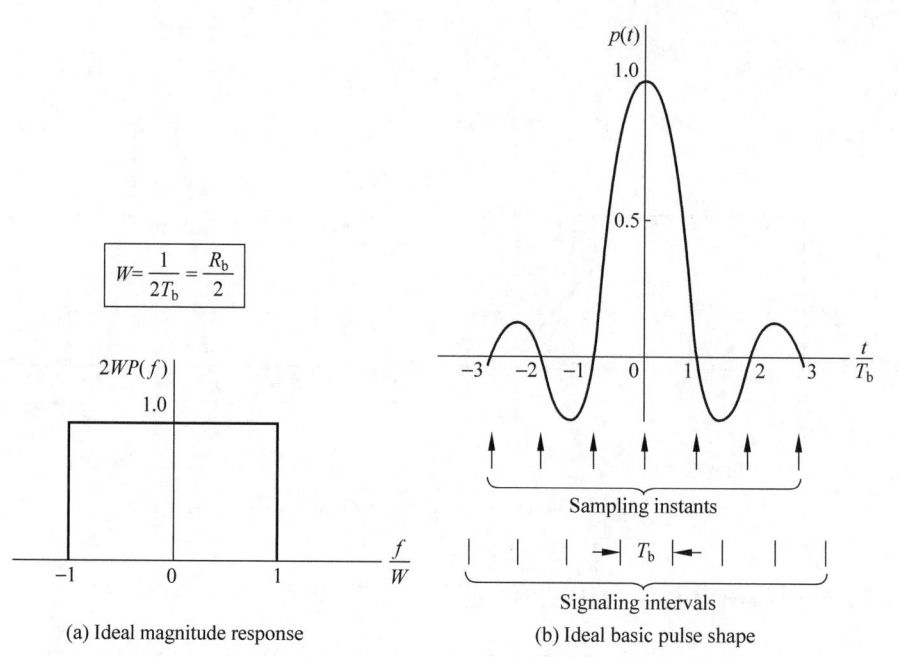

$$W = \frac{1}{2T_b} = \frac{R_b}{2}$$

(a) Ideal magnitude response　　(b) Ideal basic pulse shape

Figure 6.4.2　The ideal low-pass transmission characteristic

tail of $p(t)$ has large fluctuation and lasts a very long time(尾巴衰减慢且持续时间长).

If the transfer function $P(f)$ is a real function and has odd symmetry at $f=\omega$, equation(6.4.2) can be satisfied.

2. Raised cosine spectrum

A raised cosine spectrum satisfies the odd symmetry.

$$P(\omega)=\begin{cases} T_b, & 0 \leqslant |\omega| < \dfrac{(1-\alpha)\pi}{T_b} \\[2mm] \dfrac{T_b}{2}\left[1+\sin\dfrac{T_b}{2\alpha}\left(\dfrac{\pi}{T_b}-\omega\right)\right], & \dfrac{(1-\alpha)\pi}{T_b} \leqslant |\omega| \leqslant \dfrac{(1+\alpha)\pi}{T_b} \\[2mm] 0, & |\omega| \geqslant \dfrac{(1+\alpha)\pi}{T_b} \end{cases} \quad (6.4.11)$$

And corresponding $p(t)$ is

$$p(t)=\frac{\sin\pi t/T_b}{\pi t/T_b}\frac{\cos\alpha\pi t/T_b}{1-4\alpha^2 t^2/T_b^2} \qquad (6.4.12)$$

α is called the roll-off factor(滚降系数).

$$B_T = 2\omega - f_1 = \omega(1+\alpha) \quad \text{Transmission bandwidth}$$

$$\eta = \frac{R_B}{B} = \frac{2}{1+\alpha} \text{ (Band/Hz)} \quad \text{The maximum bandwidth efficiency}$$

When $\alpha=0$, $\eta=2$, $P(f)$ is the low-pass transmission. When $\alpha=1$, $\eta=1$, $P(f)$ is known as the full-cosine roll-off characteristic. The bandwidth is double the required bandwidth for the ideal Nyquist channel. In Figure 6.4.3(a), three values of $\alpha(0, 0.5$ and $1)$ are plotted and corresponding $p(t)$ are shown in Figure 6.4.3(b).

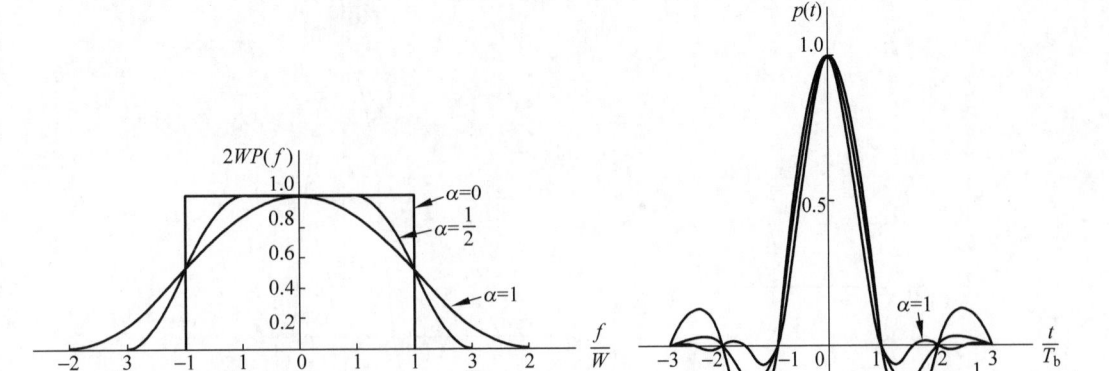

(a) Frequency response   (b) Time response

Figure 6.4.3　The raised cosine transmission characteristic

视频讲解

6.4.2　Nyquist's criterion Ⅱ

By adding intersymbol interference to transmitted signal in a controlled manner，it is possible to achieve a signaling rate equal to the Nyquist rate of 2ω symbols per second in a channel with a bandwidth of ω Hertz.

Such schemes are called correlative-level coding or partial-response signaling schemes.

The particular form of correlative-level — class Ⅰ partial response is shown in Figure 6.4.4.

Figure 6.4.4　Duobinary signaling system

Assume transfer function $H(f)$ of a baseband transmission system has ideal rectangular shape. We consider two unit impulses with space time T as the input terminal，then the output waveform should be superposition of these two $\sin x/x$ waveforms. Let the superimposed waveform be $g(t)$，then its expression will be

$$g(t) = \frac{\sin\dfrac{\pi}{T_b}\left(t+\dfrac{T_b}{2}\right)}{\dfrac{\pi}{T_b}\left(t+\dfrac{T_b}{2}\right)} + \frac{\sin\dfrac{\pi}{T_b}\left(t-\dfrac{T_b}{2}\right)}{\dfrac{\pi}{T_b}\left(t-\dfrac{T_b}{2}\right)} = \frac{4}{\pi}\left[\frac{\cos\pi t/T_b}{1-4t^2/T_b^2}\right] \qquad (6.4.13)$$

The waveform of $g(t)$ is plotted in Figure 6.4.5.

$g(t)$ decreases along with the increase of t^2. It attenuates more rapidly than sinc function.

The Fourier transform of $g(t)$ and frequency efficiency are

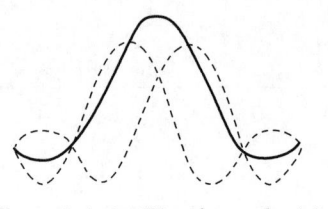

Figure 6.4.5　Waveform of $g(t)$

$$G(\omega) = \begin{cases} 2T_b \cos \dfrac{\omega T_b}{2}, & |\omega| \leqslant \dfrac{\pi}{T_b} \\ 0, & |\omega| > \dfrac{\pi}{T_b} \end{cases} \quad \text{and} \quad \eta = \dfrac{R_B}{\omega} = \dfrac{\dfrac{1}{T_b}}{\dfrac{1}{2T_b}} = 2(\text{Band/Hz})$$

This partial response system can achieve:

(1) The tails decay rapidly;

(2) The ideal frequency efficiency $\eta = 2$.

According to Figure 6.4.4, we may express duobinary coder output c_k as the sum of the present input pulse a_k and its previous value a_{k-1}

$$c_k = a_k + a_{k-1} \qquad (6.4.14)$$

where $a_k = \pm 1$, the possible values of c_k are $+2$, 0, and -2.

Let $\hat{a}_k$ represents the estimate of the original pulse a_k at the receiver at time $t = kT_s$. Then we get

$$\hat{a}_k = c_k - \hat{a}_{k-1}$$

It is apparent that if c_k is received without error and if also the previous estimate $\hat{a}_{k-1}$ is a correct decision, then $\hat{a}_k$ will be correct. But once fault decision happens, the error will be propagated, and the following received symbols will be influenced.

A rather practical partial response system is given as following Figure 6.4.6.

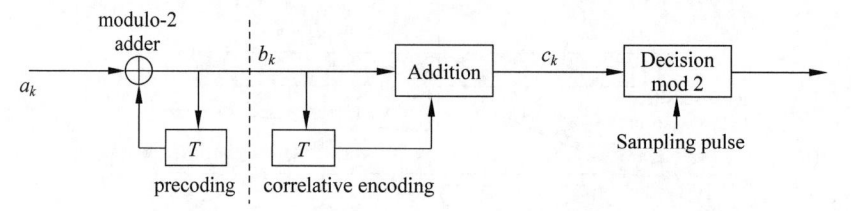

Figure 6.4.6　Block diagram of first partial response system

The precoding rule is

$$a_k = b_k \oplus b_{k-1} \qquad (6.4.15)$$

where $\oplus$ is the addition mod 2.

The correlative encoding rule is

$$c_k = b_k + b_{k-1} \qquad (6.4.16)$$

The operation of mod 2 on c_k is

$$[c_k]_{\text{mod2}} = [b_k + b_{k-1}]_{\text{mod2}} = b_k \oplus b_{k-1} = a_k \qquad (6.4.17)$$

So

$$a_k = [c_k]_{\bmod 2}$$

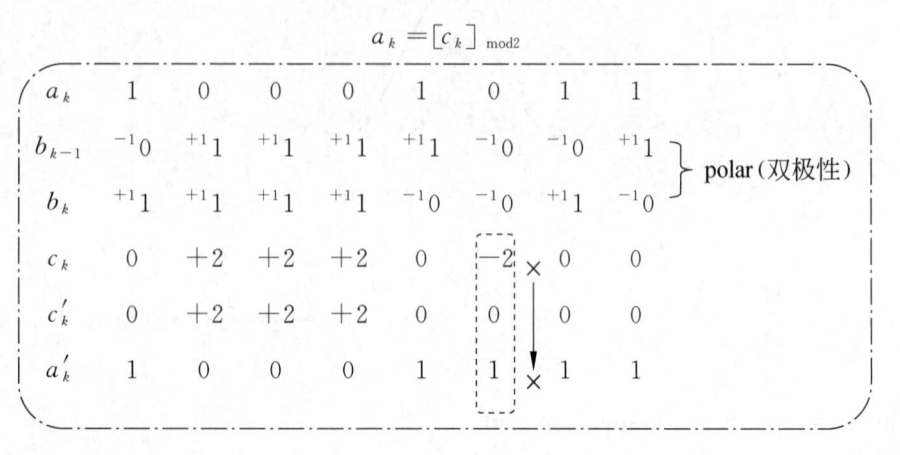

The decision rule is

$$c_k = \begin{cases} 0 & \text{say symbol is 1} \\ \pm 2 & \text{say symbol is 0} \end{cases} \tag{6.4.18}$$

Therefore, the error propagation can not occur in the detector.

6.5 Eye pattern

An experimental tool for such an evaluation in an insightful manner is the so-called eye pattern.

An eye pattern provides a great deal of useful information about the performance of a data transmission system, as described in Figure 6.5.1.

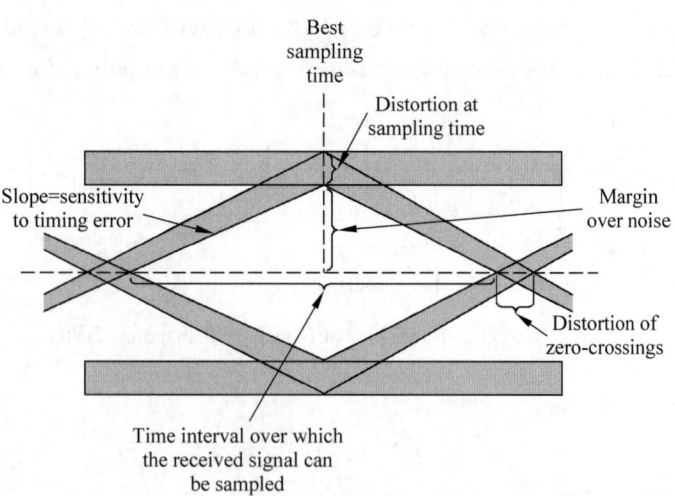

Figure 6.5.1 Interpretation of the eye pattern

(1) The width of the eye opening defines the time interval over which the received signal can be sampled without error from ISI. It is apparent that the best time for sampling is the instant of time at which the eye is open the widest.

（2）The sensitivity of the system to timing error is determined by the slope.

（3）The height of the eye opening defines the noise margin of the system.

Summary and discussion

When the digital signal is generated, it needs to be transmitted in the channel. In this chapter, we discussed the baseband case, including the baseband signal and baseband transmission system, and we will discuss the pass band in Chapter 7.

First, the baseband signal should be processed before transmission. The purpose is to let the signal match the characteristics of the channel. There are five important line codes: unipolar nonreturn-to-zero(NRZ) waveform, polar nonreturn-to-zero waveform, unipolar return-to-zero (RZ) waveform, polar return-to-zero waveform and differential encoding. These waveforms are not suitable for transmission because of the D. C components and very low frequency components. Many transmission symbol code types are designed, such as AMI, HDB_3, biphase code etc.

Second, based on the baseband transmission model, there are two important issues: noise and inter-symbol interference (ISI), which effect the signal. In order to reduce the noise, the matched filter is designed to maximize the output SNR. To eliminate or decrease the ISI, the transmission characteristic of baseband system should satisfy the Nyquist criterion. The frequency-domain condition of Nyquist criterion I is:

$$\sum_i P\left(w + \frac{2\pi i}{T_b}\right) = T_b, \quad |w| \leqslant \frac{\pi}{T_b}$$

The ideal low-pass transmission characteristics can reach the ideal band efficiency value 2Band/Hz, but can't be physically achieved. The tail has large fluctuations and lasts a very long time. Raised cosine spectrum can be realized and has fast tail, but the maximum band efficiency value is only 1Band/Hz.

Nyquist criterion II can resolve these problems: by adding inter symbol interference to transmitted signal in a controlled manner, it is possible to achieve a signaling rate equal to the Nyquist rate of 2ω symbols per second in a channel of bandwidth ω. This is also called the correlative-level coding or partial-response signaling schemes, which can achieve the ideal band efficiency value and has a fast attenuation tail. In practice, the equalizer composed of a transversal filter is used to cancel or decrease the ISI.

At last, the eye pattern is given and introduced as an experimental tool.

Homework

6.1　Suppose the information symbol is 1 1 0 0 1 0 0 0 1 1 1 0, draw the waveform of unipolar NRZ, unipolar RZ, polar NRZ, polar RZ and differential encoding.

6.2　Suppose the message data stream is 1 0 1 0 0 0 0 0 1 1 0 0 0 0 1 1, try to get the AMI and HDB$_3$ encoding.

6.3　The baseband transmission transfer function is $H(\omega)$. If the transmission rate is $\dfrac{2}{T_s}$Baud, find which one (in the following Figure 6.1) can satisfy the no ISI condition.

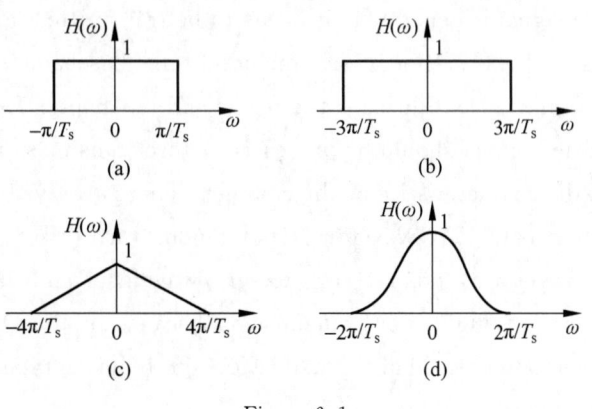

Figure 6.1

Terminologies

matched filter	匹配滤波	Polar NRZ	双极性不归零
bit error rate(BER)	误码率	Unipolar RZ	单极性归零
Nyquist's criterion	奈奎斯特准则	Polar RZ	双极性归零
partial response system	部分响应系统	Differential encoding	差分编码
intersymbol interference	码间串扰	stationary wave $v_c(t)$	稳态波
line codes	线路码	alternating wave $u_c(t)$	交变波
Unipolar NRZ	单极性不归零		

Passband data transmission

（digital passband transmission）

Mind map:

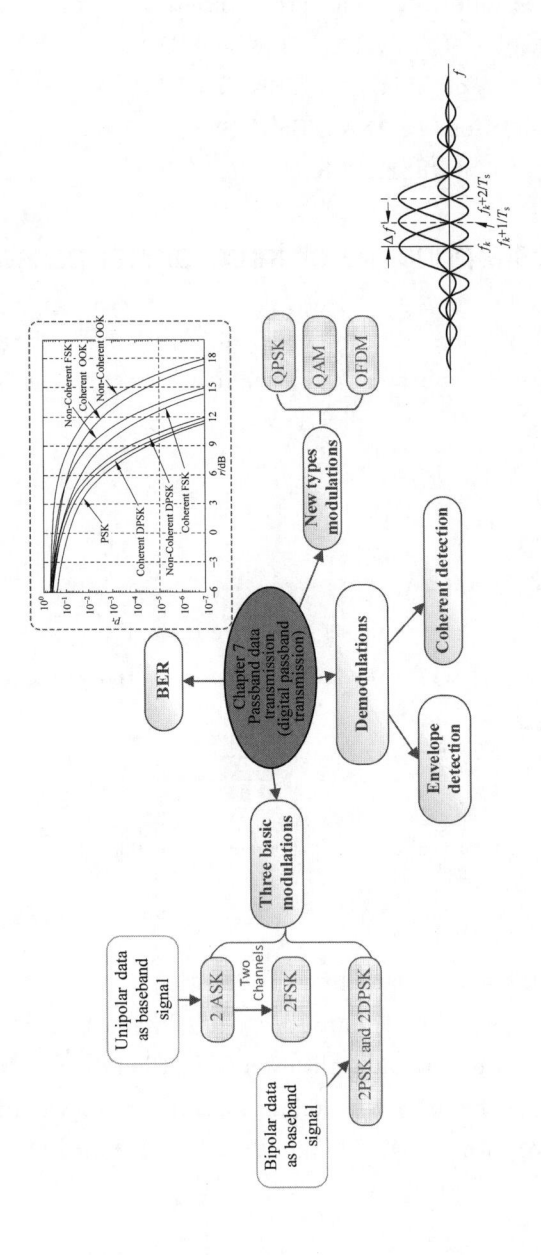

Mind map for

Chapter 7

In most cases, the baseband signal should modulate a carrier wave with a very high frequency, so that the baseband signal is shifted to a high enough frequency and can be transmitted from the antenna. This process is the digital passband transmission.

Passband modulation usually needs a sinusoidal wave as a carrier. There are three parameters of sinusoidal carrier. For binary baseband digital signal, there are three modulations：

ASK（幅移键控：Amplitude shift keying）

FSK（频移键控：Frequency shift keying）

PSK（相移键控：Phase shift keying）

In this chapter, the following topics are covered：

• The basic principles of ASK,FSK,PSK (DPSK).

• Coherent detections of ASK,FSK,PSK (DPSK).

• Noncoherent detections of ASK,FSK,DPSK.

• Bit error rates of ASK,FSK,PSK (DPSK).

视频讲解

7.1　The basic principle of three digital passband modulation

In this part, we discuss the principle of ASK, FSK, PSK and compare their differences. The often used sinusoidal carrier is usually expressed as

$$c(t) = A\cos(\omega_c t + \theta) \tag{7.1.1}$$

7.1.1　ASK

1. Basic principle

The general expression of 2ASK is

$$e_{2ASK}(t) = s(t)\cos\omega_c t \tag{7.1.2}$$

where $s(t) = \sum_n a_n g(t - nT_s)$, $g(t)$ is a rectangular pulse, as shown in Figure 7.1.1.

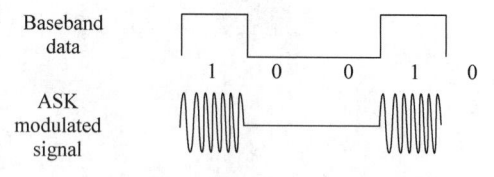

Figure 7.1.1　The waveform of ASK

There are two methods for generating the 2ASK.

The first method is by using a multiplier（乘法器）, as shown in Figure 7.1.2(a). The output signal is given by the multiplication of baseband signal $s(t)$ and carrier $c(t)$.

The second method is by using a switching circuit, as shown in Figure 7.1.2(b). The switch（开关）is controlled by $s(t)$. This method is also called OOK (on-off) keying（通断键控）.

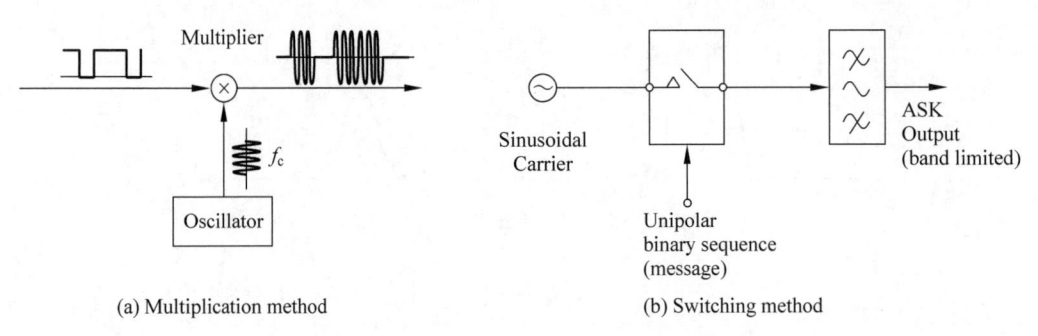

(a) Multiplication method　　　　　　(b) Switching method

Figure 7.1.2　2ASK/OOK signal modulators

In the receiver, there are two demodulation methods for ASK, i. e. the envelope detector and coherent demodulation method, which are respectively shown in Figure 7.1.3.

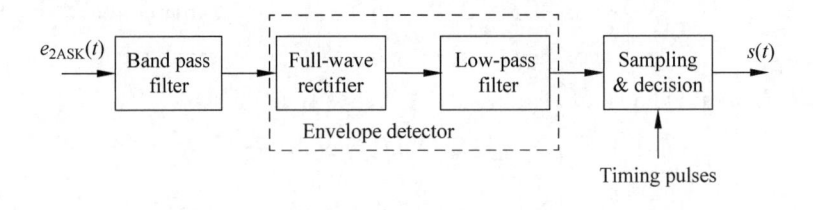

(a) Non-coherent demodulation

(b) Coherent demodulation

Figure 7.1.3　The demodulations of ASK

2. PSD

According to

$$e_{2\text{ASK}}(t) = s(t)\cos\omega_c t \tag{7.1.3}$$

where $s(t)$ is unipolar rectangular pulses. Its PSD is:

$$P_{2\text{ASK}}(f) = \frac{1}{4}\left[P_s(f+f_c) + P_s(f-f_c)\right] \tag{7.1.4}$$

where P_s is the power spectral density of $s(t)$.

Consider $P(0) = P(1) = 1/2$, we can get the PSD of 2ASK:

$$P_{2\text{ASK}}(f) = \frac{T_s}{16}\left[\left|\frac{\sin\pi(f+f_c)T_s}{\pi(f+f_c)T_s}\right|^2 + \left|\frac{\sin\pi(f-f_c)T_s}{\pi(f-f_c)T_s}\right|^2 + \frac{1}{16}\left[\delta(f+f_c) + \delta(f-f_c)\right]\right] \tag{7.1.5}$$

where $f_s = 1/T_s$ is the signal symbol rate (band).

From the following Figure 7.1.4, the wideband of 2ASK is $B_{2\text{ASK}} = 2f_s$.

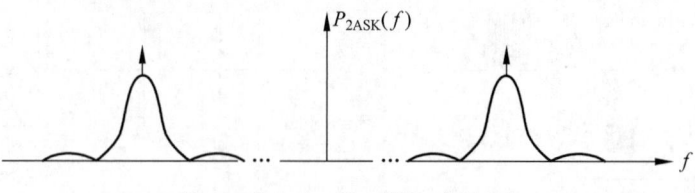

Figure 7.1.4   The PSD of 2ASK

7.1.2   FSK

1. Basic principle

The symbols 1 and 0 of 2FSK are respectively transmitted by two sinusoidal with different frequencies.

$$e_{2FSK} = \begin{cases} A\cos(\omega_1 t + \varphi_n) & \text{when } 1 \text{ is transmitted} \\ A\cos(\omega_2 t + \varphi_n) & \text{when } 0 \text{ is transmitted} \end{cases} \tag{7.1.6}$$

2FSK is the equivalence of two 2ASK superposition(叠加).

$$e_{2FSK} = \begin{cases} \sum_n a_n g(t - nT_s)\cos\omega_1 t \to ASK1 \\ \sum_n \bar{a}_n g(t - nT_s)\cos\omega_2 t \to ASK2 \end{cases} \tag{7.1.7}$$

The generation method of 2FSK is given in Figure 7.1.5, where $\tilde{a}_n$ is the inverse code（反码）of input data a_n.

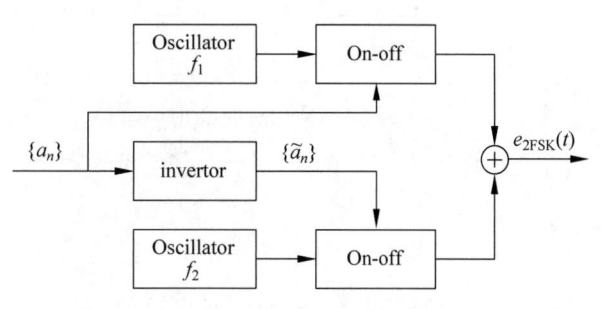

Figure 7.1.5   Generation method of 2FSK

The receiving of the FSK signal is also classified into coherent and non-coherent receivings.

The coherent receiving is equivalent to two parallel ASK demodulations, as shown in Figure 7.1.6.

If the signal in the upper branch is larger, then the received decision is 1, if the signal in the lower branch is larger, then the received decision is 0. The coherent carriers ($\cos\omega_1 t$ and $\cos\omega_2 t$) should be extracted from the received signal, so it will increase the complexity of the receiver.

The method of non-coherent receiver is given in Figure 7.1.7.

The rule of envelope detection is the same as the decision rule for the coherent receiving method.

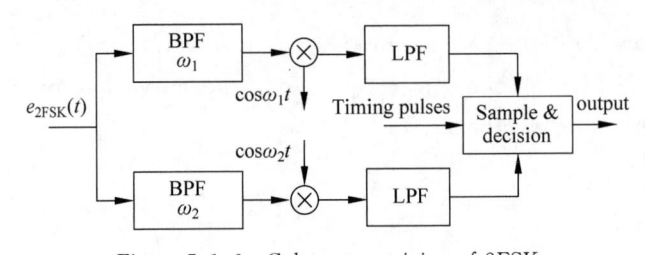

Figure 7.1.6　Coherent receiving of 2FSK

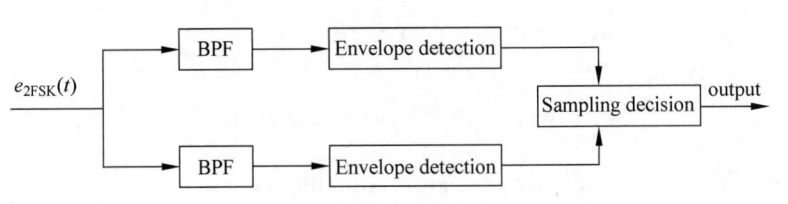

Figure 7.1.7　Envelope detection of 2FSK

Another method is the zero crossed detection, which is given in Figure 7.1.8.

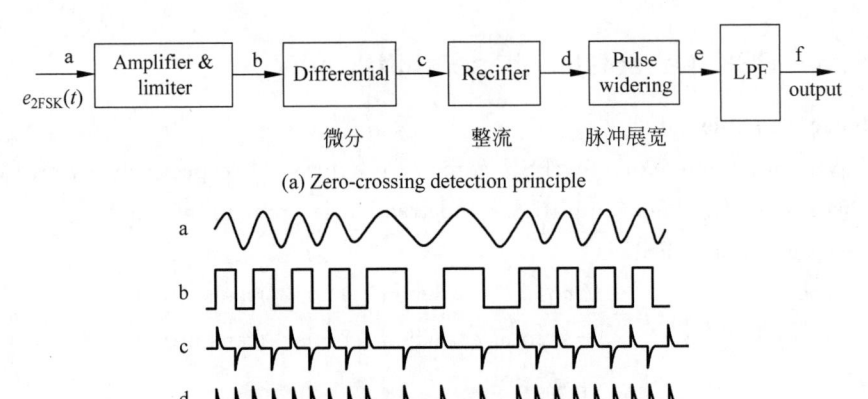

(a) Zero-crossing detection principle

(b) The waveform of each point

Figure 7.1.8　The principle and waveform of zero-crossing detection

2FSK signal is extensively used in digital communication. ITU-T recommends that 2FSK system to be used at bit rate 1200b/s and lower.

2. PSD

The PSD of 2FSK should be the sum of the PSD of two 2ASK with different frequencies. $(P(0)=P(1)=1/2)$

$$P_{2FSK}(f) = \frac{1}{16}\left[\left|\frac{\sin\pi(f+f_1)T_s}{\pi(f+f_1)T_s}\right|^2 + \left|\frac{\sin\pi(f-f_1)T_s}{\pi(f-f_1)T_s}\right|^2 + \left|\frac{\sin\pi(f+f_2)T_s}{\pi(f+f_2)T_s}\right|^2 + \right.$$

$$\left. \left|\frac{\sin\pi(f-f_2)T_s}{\pi(f-f_2)T_s}\right|^2\right] + \frac{1}{16}\left[\delta(f+f_1)+\delta(f-f_1)+\delta(f+f_2)+\delta(f-f_2)\right]$$

$$(7.1.8)$$

As can be seen from the above equation （7.1.8）, the PSD curve of 2FSK can be plotted in Figure 7.1.9. When $|f_2-f_1|<f_s$, the curve has only one peak; when $|f_2-f_1|\geqslant f_s$, there are two peaks.

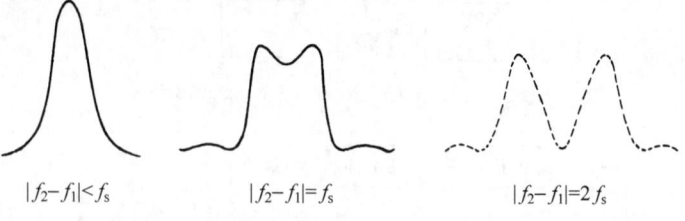

$|f_2-f_1|<f_s$ $|f_2-f_1|=f_s$ $|f_2-f_1|=2f_s$

Figure 7.1.9　The PSD of 2FSK

The bandwidth of the 2FSK signal approximately equals to

$$\Delta f = |f_2 - f_1| + 2f_s \tag{7.1.9}$$

where $f_s = \dfrac{1}{T_s}$.

7.1.3　PSK（absolute phase shift keying）

1. Basic principle

The symbols 0 and 1 of the 2PSK signal are expressed respectively by two different initial phases 0 and π. Their amplitudes and frequencies remain unchanged.

The expression of 2PSK is

$$e_{2PSK}(t) = \begin{cases} A\cos\omega_c t & \text{when } 0 \text{ is transmitted} \\ -A\cos\omega_c t & \text{when } 1 \text{ is transmitted} \end{cases} \tag{7.1.10}$$

or written as

$$e_{2PSK}(t) = s(t)\cos\omega_c t \tag{7.1.11}$$

when $s(t) = \sum_n a_n g(t - nT_s)$, a_n is the Polar NRZ(双极性不归零)rectangular pulse.

The 2PSK modulator is illustrated in Figure 7.1.10, including the waveforms.

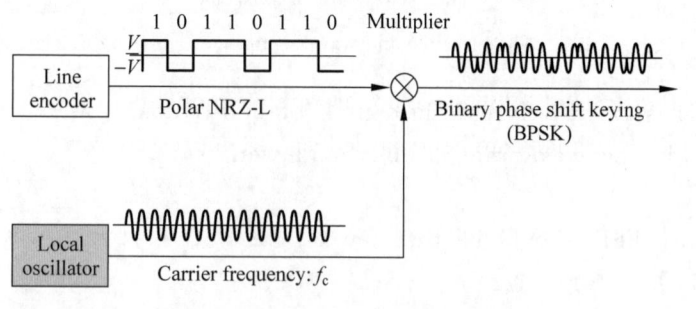

Figure 7.1.10　2PSK modulator

The demodulation method of 2PSK signal is the coherent receiving, as shown in Figure 7.1.11.

And waveform of each point is shown in Figure 7.1.12.

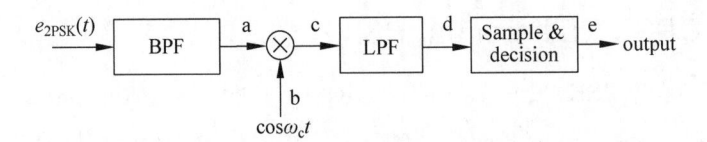

Figure 7. 1. 11　2PSK coherent demodulation

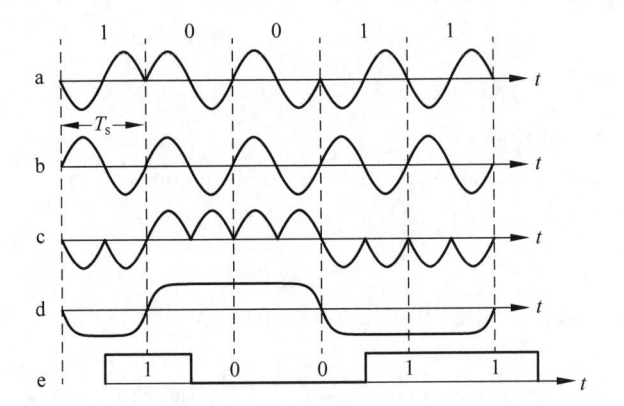

Figure 7. 1. 12　The waveform of each point

There may be a phase ambiguity（相位模糊）of $\cos\omega_c t$ at the receiver, which will cause phase 0 and π inversing. The symbols 1 and 0 are inversed, and error decision is made. In order to overcome this shortcoming, the DPSK system is usually adopted in practice, which will be discussed in Section 7. 1. 4.

2. PSD

For 2PSK signal $g_1(t)=-g_2(t)$, and their spectra $G_1(f)=-G_2(f)$. When $P(0)=P(1)=1/2$,

$$P_{2PSK}(f) = \frac{1}{4}f_c \left[\,|G_1(f+f_c)|^2 + |G_1(f-f_c)|^2\,\right]$$

$$= \frac{T_s}{4}\left[\left|\frac{\sin\pi(f+f_c)T_s}{\pi(f+f_c)T_s}\right|^2 + \left|\frac{\sin\pi(f-f_c)T_s}{\pi(f-f_c)T_s}\right|^2\right] \qquad (7.1.12)$$

Compared equation(7. 1. 6) with equation(7. 1. 9), and Figure 7. 1. 4 with Figure 7. 1. 13, the difference is that PSD of 2PSK have no discrete components $\delta(f+f_c)$ and $\delta(f-f_c)$. Therefore, the 2PSK signal can be regarded as s suppressed carrier DSB ASK. The bandwidth of 2PSK is $B_{2PSK}=2f_s$.

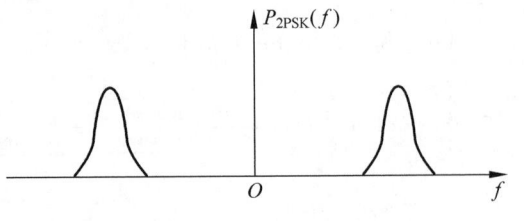

Figure 7. 1. 13　PSD of 2PSK signal

7.1.4 DPSK（differential PSK）

The DPSK（差分 PSK）utilizes the relative value of the carrier phases of the adjacent symbols to express baseband signals 0 and 1. θ is used to express the initial phase of the carrier.

$\Delta\theta$ is the phase difference of the current symbol and the previous symbol，the rule is

$$\begin{cases} \Delta\theta = 0 & \text{when} \quad 0 \quad \text{is} \quad \text{transmitted} \\ \Delta\theta = \pi & \text{when} \quad 1 \quad \text{is} \quad \text{transmitted} \end{cases} \tag{7.1.13}$$

For example：（Initial phase is 0）

Baseband sequence a_k： 1 1 1 0 0 1 1 0   Absolute code

DPSK   (0) b_k： 1 0 1 1 1 0 1 1   Relative code

$$\boxed{b_k = a_k \oplus b_{k-1}} \tag{7.1.14}$$

The generation diagram of 2DPSK is given in Figure 7.1.14.

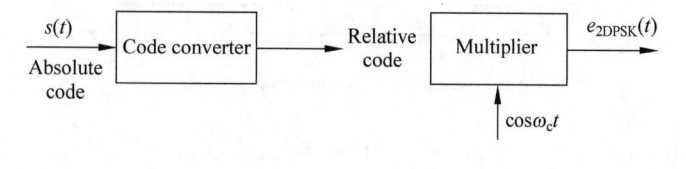

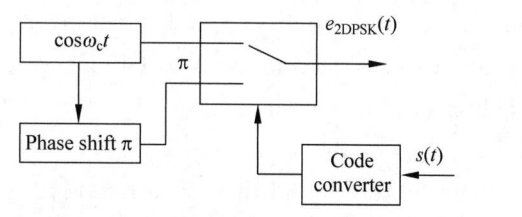

Figure 7.1.14   Generation method of 2DPSK

There are mainly two methods for demodulation of DPSK signal. The first method is direct comparison of the phases of adjacent symbols，as shown in Figure 7.1.15.

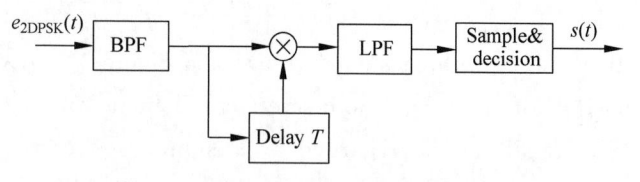

Figure 7.1.15   Direct method of DPSK demodulation

The second method is the coherent demodulation，as shown in Figure 7.1.16.

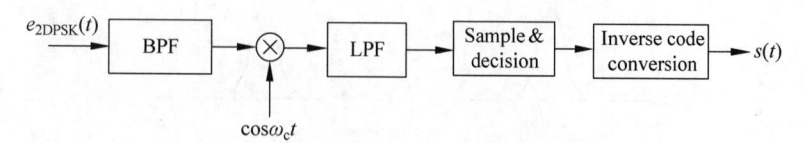

Figure 7.1.16   The coherent demodulation of 2DPSK

7.2　Anti-noise performance of digital passband modulation system

7.2.1　Bit error rate of ASK

1. Coherent modulation

For a 2ASK system, the sketch map of bit error rate is shown in Figure 7.2.1, which is similar with the unipolar baseband system.

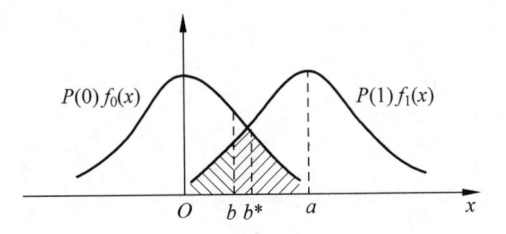

Figure 7.2.1　Bit error rate of 2ASK

The additive white Gaussian noise in the channel is considered.

$$n(t) = n_c(t)\cos\omega_c t - n_s(t)\sin\omega_c t \tag{7.2.1}$$

After multiplied by the local carrier and passing through the LPF, the received signal is $y(t)$, its voltage at the sampling & decision point can be expressed by

$$x(t) = \begin{cases} A + n_c(t) & \text{when } 1 \text{ is transmitted} \\ n_c(t) & \text{when } 0 \text{ is transmitted} \end{cases} \tag{7.2.2}$$

It is similar with the unipolar baseband system.

When $P(1) = P(0)$,

$$V_d^* = \frac{A}{2} \tag{7.2.3}$$

and

$$P_e = \frac{1}{2}\text{erfc}\sqrt{\frac{r}{4}} \tag{7.2.4}$$

where $r = \dfrac{a^2}{2\delta_n^2}$, r is the input signal-to-noise ratio of demodulation.

2. Envelope detection

In the envelope system, the conditional probability distribution functions are written in equation (7.2.5) and equation (7.2.6), the curves are shown in Figure 7.2.2.

$$f_1(V) = \frac{V}{\delta_n^2}I_0\left(\frac{aV}{\delta_n^2}\right)e^{-(V^2+a^2)/2\delta_n^2} \tag{7.2.5}$$

$$f_0(V) = \frac{V}{\delta_n^2}e^{-V^2/2\delta_n^2} \tag{7.2.6}$$

$f_1(V)$ obeys the generalized Rayleigh distribution, $f_0(V)$ obeys the Rayleigh distribution.

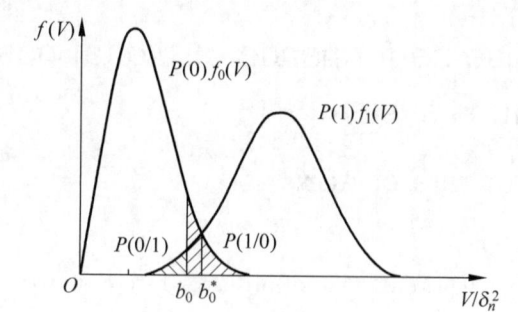

Figure 7.2.2 The geometric representation of ASK P_e for envelope detection

When $P(0)=P(1)$,

$$V^* = \frac{A}{2} \qquad\qquad (7.2.7)$$

$$P_e = \frac{1}{4}\mathrm{erfc}\sqrt{\frac{r}{2}} + \frac{1}{2}e^{-r/4} \qquad\qquad (7.2.8)$$

When $r \to \infty$, $P_e = \frac{1}{2}e^{-r/4}$.

Example 7.2.1: for an ASK signal transmission system, assume the symbol rate is $R_b = 4.8 \times 10^6$ Baud, the amplitude of the received signal is $A - 1\mathrm{mV}$ and the single-side PSD of the Gaussian noise is $n_0 = 2 \times 10^{-15}$ W/Hz. Find:

(1) the optimum symbol error probability when the envelope detection is used;

(2) the optimum symbol error probability when the coherent demodulation is used.

Solution: The bandwidth of 2ASK is $2/T$ Hz. According to the signal rate, the optimum bandwidth of the bandpass filter in the receiver should be selected as

$$B = 2/T = 2R_b = 9.6 \times 10^6 (\mathrm{Hz})$$

Therefore, the mean power of the output noise of the bandpass filter is

$$\delta_n^2 = n_0 B = 1.92 \times 10^{-8} (\mathrm{W})$$

The output signal to noise ratio is

$$r = \frac{A^2}{2\delta_n^2} = \frac{10^{-6}}{2 \times 1.92 \times 10^{-8}} \approx 26 \gg 1$$

For the envelope detection

$$P_e = \frac{1}{2}e^{-r/4} = \frac{1}{2}e^{-6.5} = 7.5 \times 10^{-4}$$

For the coherent demodulation

$$P_e = \frac{1}{\sqrt{\pi r}}e^{-r/4} = \frac{1}{\sqrt{3.14 \times 26}}e^{-6.5} = 1.66 \times 10^{-4}$$

7.2.2 Bit error rate of 2FSK

1. Coherent demodulation

According to the block diagram of 2FSK coherent demodulation.

Suppose symbol 1 is transmitted, the two received voltages after passing the two band-pass filters can be written as

$$y_1(t) = [A + n_c(t)]\cos\omega_1 t - n_{1s}(t)\sin\omega_1 t$$
$$y_2(t) = n_{2c}(t)\cos\omega_2(t) - n_{2s}(t)\sin\omega_c t \tag{7.2.9}$$

They are multiplied by the local carrier, and pass through the LPF, so the inputs of decision-making circuit are

$$V_1(t) = A + n_{1c}(t)$$
$$V_2(t) = n_{2c}(t) \tag{7.2.10}$$

Because $P(0/1) = P(V_1 < V_2)$ and $P(1/0) = P(V_1 > V_2)$

When $P(0) = P(1)$,

$$P_e = \frac{1}{2}\text{erfc}\sqrt{\frac{r}{2}} \tag{7.2.11}$$

When $r \gg 1$,

$$P_e \approx \frac{1}{\sqrt{2\pi r}}e^{-\frac{r}{2}} \tag{7.2.12}$$

2. Envelope detection

When symbol 1 is transmitted, the two voltages at the sampling & decision device input are

$$V_1(t) = \sqrt{[A + n_{c1}(t)]^2 + n_{s1}^2(t)} \text{ and } V_2(t) = \sqrt{n_{c2}^2(t) + n_{s2}^2(t)}$$
$$P_{e1} = P(V_1 < V_2) \text{ and } P_{e2} = P(V_1 > V_2)$$

so the bit error rate is

$$P_e = \frac{1}{2}e^{-\frac{r}{2}} \tag{7.2.13}$$

Example 7.2.2: assume there is a 2FSK transmission system, its transmission bandwidth is 2400Hz; the two carrier frequencies of the 2FSK signal are respectively $f_0 = 980$Hz, $f_1 = 1580$Hz, the symbol rate is $R_B = 300$Baud, and the signal to noise ratio at the input of the receiver is 6dB.

Find: (1) the bandwidth of this FSK signal;

(2) the symbol error probability when the envelope detection is used;

(3) the symbol error probability during the coherent demodulation.

Solution: (1) The signal bandwidth is

$$\Delta f = |f_1 - f_0| + 2f_c = 1580 - 980 + 2 \times 300 = 1200(\text{Hz})$$

(2) The symbol error probability of the envelope detection is

$$P_e = \frac{1}{2}e^{-\frac{r}{2}}$$

which is decided by the signal to noise ratio r. Now the given input signal to noise ratio is equal to 6dB, but after passing the bandpass filter the noise will further limited, and the signal to noise will be increased. The bandwidth of the bandpass filter should be equal to: $B = 2R_b = 600$Hz. The bandwidth ratio of the input and output of the bandpass filter equals

$2400/600=4$. So the noise power will be reduced to $1/4$, i. e. The signal to noise power ratio will be increased by 4 times. Thus the signal to noise ratio at the receiver is $r=4\times4=16$. So

$$P_e = \frac{1}{2}e^{-\frac{r}{2}} = \frac{1}{2}e^{-8} = 1.7\times10^{-4}$$

（3）The symbol error probability of the coherent demodulation is

$$P_e = \frac{1}{2}\text{erfc}\sqrt{\frac{r}{2}} = \frac{1}{2}\left[1-\text{erf}\sqrt{\frac{r}{2}}\right] = \frac{1}{2}[1-\text{erf}2.8284] = 3.5\times10^{-5}$$

and the approximation is

$$P_e \approx \frac{1}{\sqrt{2\pi r}}e^{-\frac{r}{2}} = 3.39\times10^{-5}$$

7.2.3 Bit error rate of PSK

For 2PSK system，the sketch map of bit error rate is shown in Figure 7.2.3，which is similar with the polar baseband system.

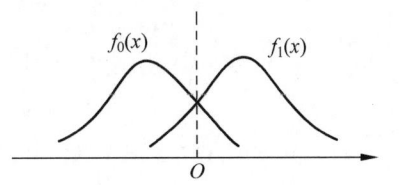

Figure 7.2.3　Bit error rate of 2PSK

1. Coherent demodulation of 2PSK

The input voltage of the sampling and decision device can be written as

$$x(t) = \begin{cases} a + n_c(t) & \text{when } 1 \text{ is transmitted} \\ -a + n_c(t) & \text{when } 0 \text{ is transmitted} \end{cases} \tag{7.2.14}$$

Because

$$n_c(t) \sim N(0, \delta_n^2)$$

So

$$f_1(x) = \frac{1}{\sqrt{2\pi}\delta_n}\exp\left[-\frac{(x-a)^2}{2\delta_n^2}\right] \quad \text{when } 1 \text{ is transmitted}$$

$$f_0(x) = \frac{1}{\sqrt{2\pi}\delta_n}\exp\left[-\frac{(x+a)^2}{2\delta_n^2}\right] \quad \text{when } 0 \text{ is transmitted}$$

The best decision threshold is

$$V_d^* = 0$$

Then the bit error rate is

$$P_e = \frac{1}{2}\text{erfc}\sqrt{r}$$

When $r \gg 0$ or $r \rightarrow \infty$

$$P_e \approx \frac{1}{2\sqrt{\pi r}} \mathrm{e}^{-r} \qquad (7.2.15)$$

2. Phase comparison method of DPSK

$P(0/1) = P\{x < 0\}$

$$= P\left\{\frac{1}{2}[(a+n_{1c})(a+n_{2c})+(n_{1s}n_{2s})] < 0\right\}$$

$$= P\{[(2a+n_{1c}+n_{2c})^2+(n_{1s}+n_{2s})^2-(n_{1c}-n_{2c})^2-(n_{1s}-n_{2s})^2] < 0\}$$

Let

$$R_1^2 = (2a+n_{1c}+n_{2c})^2+(n_{1s}+n_{2s})^2 \text{ and } R_2^2 = (n_{1c}-n_{2c})^2+(n_{1s}-n_{2s})^2$$

$$P(0/1) = P\{R_1 < R_2\}, \quad f(R_1) = \frac{R_1}{2\delta_n^2} I_0\left(\frac{aR_1}{\delta_n^2}\right) \mathrm{e}^{-(R_1^2+4a^2)/4\delta_n^2}$$

$$P(0/1) = \frac{1}{2} \mathrm{e}^{-r}, \quad f(R_2) = \frac{R_2}{2\delta_n^2} \mathrm{e}^{-R_2^2/4\delta_n^2}$$

So

$$P(1/0) = \frac{1}{2} \mathrm{e}^{-r}$$

When

$$P(0) = P(1) = \frac{1}{2}, \quad P_e = \frac{1}{2} \mathrm{e}^{-r} \qquad (7.2.16)$$

3. Coherent detection of 2DPSK

The difference between PSK and DPSK is the code inverse converter, as shown in Figure 7.2.4. We only discuss the bit error rate of the code inverse converter.

Figure 7.2.4   Code inverse converter

The output probability of the code inverse converter is

$$P'_e = 2P_e = \mathrm{erfc}\sqrt{r} \qquad (7.2.17)$$

7.2.4   Performance comparison of digital keying transmission system

The symbol error rate probability equations of various binary keying systems discussed in this chapter are listed in Table 7.2.1. As can be seen from these equations, the symbol error probabilities of the coherent demodulation are all related to erfc(r) and the symbol error probabilities of the non-coherent demodulation are all related to exp($-r$).

The symbol error probability curves are plotted according to these equations are shown in Figure 7.2.5. From this figure, we can see that for the same keying mode, the symbol error probabilities of the coherent demodulation are all less than those of the non-coherent demodulations, and symbol error probability of PSK is the best.

Table 7.2.1 The bite error rate of digital passband system

Modulation	Coherent detection	Non-coherent detection
2ASK	$\frac{1}{2}\mathrm{erfc}\left(\sqrt{\frac{r}{4}}\right)$	$\frac{1}{2}e^{-r/4}$
2FSK	$\frac{1}{2}\mathrm{erfc}\left(\sqrt{\frac{r}{2}}\right)$	$\frac{1}{2}e^{-r/2}$
2PSK	$\frac{1}{2}\mathrm{erfc}(\sqrt{r})$	—
2DPSK	$\mathrm{erfc}(\sqrt{r})$	$\frac{1}{2}e^{-r}$

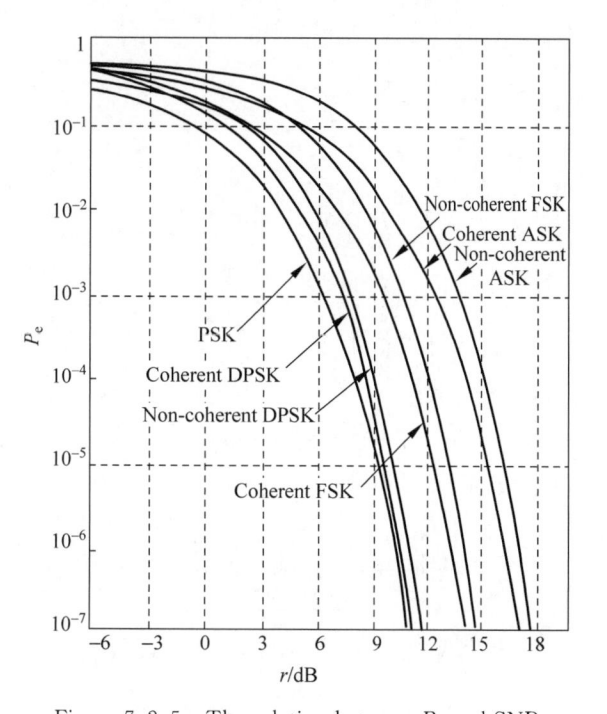

Figure 7.2.5 The relation between P_e and SNR

视频讲解

7.3 Hybrid amplitude/phase modulation schemes

In an M-ary PSK system, the in-phase and quadrature components of the modulated signal are interrelated in such a way that the envelope is constrained to remain constant. This constraint manifests itself in a circular constellation for the message points. However, if the constraint is removed, and the in-phase and quadrature components are thereby permitted to be independent, which leads to get a new modulation scheme called M-ary quadrature amplitude modulation (QAM). Both amplitude and phase modulations are involved in modulating the carrier in this scheme.

In QAM, there are two basis functions, as shown by

$$\varphi_1(t) = \sqrt{\frac{2}{T}}\cos(2\pi f_c t), \quad 0 \leqslant t \leqslant T$$

$$\varphi_2(t) = \sqrt{\frac{2}{T}}\sin(2\pi f_c t), \quad 0 \leqslant t \leqslant T$$

The transmitted M-ary QAM signal for symbol k, is defined by

$$s_k(t) = \sqrt{\frac{2E_0}{T}}a_k\cos(2\pi f_c T) - \sqrt{\frac{2E_0}{T}}b_k\sin(2\pi f_c t), \quad 0 \leqslant t \leqslant T, k = 0, +1, +2, \cdots$$

The signal $s_k(t)$ consists of two phase-quadrature carriers with each one being modulated by a set of discrete amplitudes, hence the name is quadrature amplitude modulation, as illustrated in Figure 7.3.1.

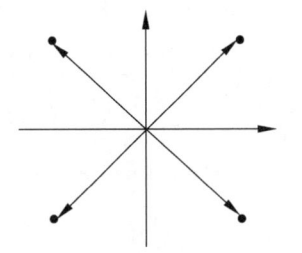

Figure 7.3.1　Signal-space diagram of 4QAM

16-QAM square constellation is as following Figure 7.3.2.

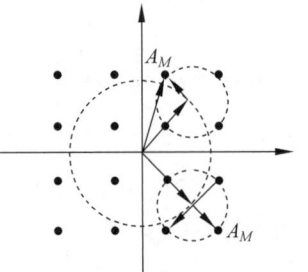

(a) Signal-space diagram of M-ary QAM for M=16;　(b) The generation principle of 16QAM
the message points in each quadrant are identified
with Gray-encoded quadbits

Figure 7.3.2　16-QAM square constellation

7.4　OFDM system

视频讲解

Orthogonal frequency division multiplexing (OFDM) is a kind of multi-carrier parallel modulation system. The main aims of this scheme are:

(1) To improve the frequency utilization and increase the transmission rate. Frequency spectra of modulated subcarriers are partially overlapped;

(2) Modulated signals are strictly orthogonal to each other in order to be completely separated in the receiver;

(3) Modulation of each subcarrier is M-ary modulation;

(4) The modulation system of each subcarrier can be different and adaptive to the variation of the channel.

OFDM has been widely used in ADSL，HDTV，DVB（digital video broadcasting），wireless local area networks，and etc. It has been the focus of research to be used in the next generation of cellular networks.

Assume there are N subchannels in an OFDM system，each subchannel uses a subcarrier：

$$x_k(t) = B_k \cos(2\pi f_k t + \varphi_k), \quad k = 0,1,2,\cdots,N-1 \qquad (7.4.1)$$

where B_k is the amplitude of the kth subcarrier，decided by the input symbols；f_k is the carrier frequency of the kth subchannel. φ_k is the initial phase of the carrier of the kth subchannel，then the sum of the N-sub signal in the system can be expressed as

$$s(t) = \sum_{k=0}^{N-1} x_k(t) = \sum_{k=0}^{N-1} B_k \cos(2\pi f_k t + \varphi_k) \qquad (7.4.2)$$

It also can be rewritten as the complex form as follows：

$$s(t) = \sum_{k=0}^{N-1} B_k e^{j2\pi f_k t + \varphi_k} \qquad (7.4.3)$$

where B_k is the complex input data of the kth subchannel.

If the frequency spacing between adjacent subcarrier are identical and equal the reciprocal of the symbol duration $\Delta f = 1/T$，and the subcarrier frequencies equal $f_k = (m+n)/2T$，$m = 1,2,\cdots,n = 1,2,\cdots$.

Then arbitrary two subcarriers in the symbol duration T are orthogonal，i. e.

$$\int_0^T \cos(2\pi f_k t + \varphi_k) \cos(2\pi f_i t + \varphi_i) \mathrm{d}t = 0 \qquad (7.4.4)$$

where $f_i = (m-n)/2T$. The hold of the orthogonal condition in equation（7.4.4）is the independent of the values of φ_k and φ_i. Therefore，this multi-subcarrier system is called OFDM system，as shown in Figure 7.4.1.

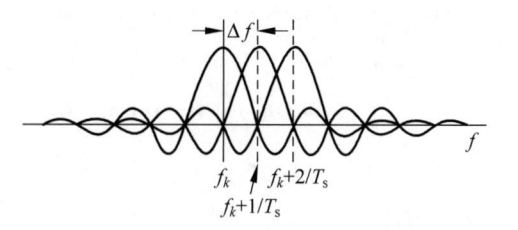

Figure 7.4.1　The spectra of multichannel subcarriers

Summary and discussion

In the digital pass band system，the carrier wave is also sine wave. We can compare this with continuous wave modulation in Chapter 4. The difference is that the baseband signal in this chapter is digital signal, not the analog signal. The three modulation parameters of carrier are also amplitude, frequency and phase, so there are three basic digital modulations: 2ASK, 2FSK and 2PSK. Since there is phase ambiguity in the 2PSK, then the 2DPSK is usually adopted in practice. Here the spectrum of modulated signal is the shift spectrum of digital baseband signal, which is more suitable for transmission. We have studied the time domain and frequency domain of baseband digital signal in chapter 5, which is also the basic theory of this chapter.

The general model of passband data transmission system should be remembered. The input signal of ASK can be regarded as unipolar signal, while the input of 2PSK can be regarded as polar signal. So the anti-noise performance can be also compared with the unipolar and polar systems and the results are similar. The different of 2FSK system is that there are two carrier wave frequencies. Compare three modulations from generation, spectrum, bandwidth, demodulation and anti-noise performance, you can understand the system more fluently(consider another form).

ASK is used almost in every digital communication link including your cell phone and cable TV. In most wireless links such as satellite TV and high definition TV broadcast channels uses two ASK links in parallel, each of 16-levels, but rotated in phase 90 degrees, thus the 16×16 combination is known as 256 QAM.

2FSK transmitter and FSK receiver implementations are simple for low data rate application. The 2FSK systems can provide high SNR and it has lower probability of error and higher immunity to noise due to constant envelope. But it uses larger bandwidth compared to other modulation techniques such as ASK and PSK. Hence it is not bandwidth efficient. The BER (bit error rate) performance in AWGN channel is worse compared to PSK modulation. In order to overcome drawbacks of 2FSK, MFSK modulation techniques with more than two frequencies have been developed.

For the PSK system, it carries data over RF signal more efficiently compared to other modulation types. Hence, it is more power efficient modulation technique compared to ASK and FSK. 2PSK system is less susceptible to errors compared to ASK modulation and occupies same bandwidth as ASK. Higher data rate of transmission can be achieved using high level of PSK modulations such as QPSK (represents 2 bits per constellation), 16-QAM (represents 4 bits per constellation) etc.

Homework

7.1   If the message signal is 101101, draw the waveform of ASK, FSK, PSK and DPSK.

7.2   Assume the rate of 1Mb/s is required to transmit data using the 2DPSK system, and the symbol error probability does not exceed 10^{-4}, also single-side power spectral density of the white Gaussian noise at the receiver input is $n_0 = 10^{-12}$ W/Hz. Find:

(1) the required received signal power for the phase comparison method;

(2) the required received signal power for the coherent detection method.

Terminologies

ASK(amplitude shift keying)   幅移键控

FSK(frequency shift keying)   频移键控

PSK(phase shift keying)   相移键控

DPSK(differential PSK)   差分相移键控

QPSK   正交相移键控

QAM   正交振幅调制

OFDM   正交频分复用

Chapter 8 | Further reading: new technologies

in communication systems

8.1 Compressive sensing (CS)

According to the sampling theory, the sampling rate is usually $5 \sim 10$ times of the maximum frequency of the message signal in practice, and sometimes there is a redundancy of data. So is there any other method to recovery the original signal with fewer samples? Recently, the compressive sensing theory has been proposed and become widely used in communication, radar and image processing systems.

8.1.1 Introduction

An early breakthrough in signal processing was the Nyquist-Shannon sampling theorem. It states that if the signal's highest frequency is less than half of the sampling rate, then the signal can be reconstructed perfectly by means of sinc interpolation. The main idea is that with prior knowledge about constraints on the signal's frequencies, fewer samples are needed to reconstruct the signal.

Around 2004, Emmanuel Candès, Justin Romberg, Terence Tao, and David Donoho proved that given knowledge about a signal's sparsity, the signal may be reconstructed with even fewer samples than the sampling theorem requires[1]. This idea is the basis of compressed sensing. Compressed sensing (also known as compressive sensing, compressive sampling, or sparse sampling) is a signal processing technique for efficiently acquiring and reconstructing a signal, by finding solutions to underdetermined linear systems. This is based on the principle that, through optimization, the sparsity of a signal can be exploited to recover it from far fewer samples than required by the Shannon-Nyquist sampling theorem. The basic assumption in CS approach is that most of the signals in real applications have a concise representation in a certain transform domain where only few of them are significant, while the rest are zero or negligible. There are two conditions under which recovery is possible. One condition is defined as signal sparsity, which requires the signal to be sparse in some domain. Another important requirement is the incoherent nature of measurements（observations）in the signal acquisition domain. Therefore，the main objective of CS is to provide an estimate of the original signal from a small number of linear incoherent measurements by exploiting the sparsity property.

8.1.2　The mathematics theory

Imagine we have a signal x , which represents a time-domain signal of length N . In a perfect world，we would simply take N measurements. However，we can only take M measurements（$M \ll N$）each time. We use a vector y to represent our M measurements. What's important to note is that one measurement doesn't necessarily correspond to a single input value. This may seem confusing at first，but in real-life systems measurements aren't always made in a single-file，linear fashion. In order to get to y ，we use $\boldsymbol{\Phi}$. Put simply：

$$y = \boldsymbol{\Phi} x \tag{8.1.1}$$

$\boldsymbol{\Phi}$ is the sensing matrix，which allows us to get from x to y （via random measurements，transformations，or a combination of the two）. To summarize，we have the following Figure 8.1.1.

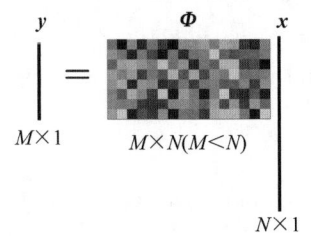

Figure 8.1.1　The illustration of equation（8.1.1）

- x —$N \times 1$ vector representing original signal（must be somehow compressible）.
- y —$M \times 1$ vector of output values.
- $\boldsymbol{\Phi}$ —$M \times N$ sensing matrix（how we get from input to output）.

x can be uniquely reconstructed by solving the following well-known convex optimization problem（凸优化问题）：

$$\text{minimize } \|x\|_1 \quad \text{subject to} \quad \boldsymbol{\Phi} x = y \tag{8.1.2}$$

where $\|x\|_1$ is ℓ_1 norm（ℓ_1 范数）. This is a NP hard problem. There are basic three type algorithms：matching pursuit（MP），basis pursuit（BP），bayesian compressive sensing（BCS）and their extending methods，which are usually used in signal processing[2]. The signal process of CS is shown in Figure 8.1.2.

$$x \rightarrow \boxed{\text{Sampling}} \xrightarrow{N} \boxed{\text{Compressing}} \xrightarrow{K} \boxed{\text{Transmission}} \rightarrow \boxed{\text{Receiver}} \xrightarrow{K} \boxed{\text{Decompressing}} \xrightarrow{N} \hat{x}$$

Figure 8.1.2　The process of compressive sensing

The comparison table between Nyquist's sampling and compressive sensing are in Table 8.1.1

Table 8.1.1 The comparison between Nyquist's sampling and compressive sensing

Comparison	Nyquist's sampling	Compressive sensing
Sampling Frequency	$\geqslant 2f_H$	$< f_H$
Recovery	Low pass filter	Convex optimization

Exercise 8.1.1: Imagine x is of length $N=8$ and has random values. We want to keep only 3 of them (let's say the 1st, 2nd and 5th). Here's what the linear algebra would look like in terms of $y = Ax$ in equation (8.3) [3].

$$\begin{bmatrix} 0.89 \\ 0.18 \\ 0.69 \end{bmatrix} = \begin{bmatrix} 1 & 0 & 0 & 0 & 0 & 0 & 0 & 0 \\ 0 & 1 & 0 & 0 & 0 & 0 & 0 & 0 \\ 0 & 0 & 0 & 0 & 1 & 0 & 0 & 0 \end{bmatrix} \begin{bmatrix} 0.89 \\ 0.18 \\ 0.03 \\ 0.37 \\ 0.69 \\ 0.54 \\ 0.11 \\ 0.90 \end{bmatrix} \tag{8.1.3}$$

In equation (8.1.3), we are clearly keeping three samples of our original values and throwing away the other 5. If x is a sparse signal (only a few useful values) then we can not afford to randomly keep a few values, or else we might lose the ones we need. This is because we do not know the location of the useful values beforehand.

$$\begin{bmatrix} 0 \\ 0.18 \\ 0 \end{bmatrix} = \begin{bmatrix} 1 & 0 & 0 & 0 & 0 & 0 & 0 & 0 \\ 0 & 1 & 0 & 0 & 0 & 0 & 0 & 0 \\ 0 & 0 & 0 & 0 & 1 & 0 & 0 & 0 \end{bmatrix} \begin{bmatrix} 0 \\ 0.18 \\ 0.03 \\ 0 \\ 0 \\ 0 \\ 0.11 \\ 0 \end{bmatrix} \tag{8.1.4}$$

If we randomly select values from our sparse signal, we lose some information. We need a different approach so we do not lose our non-zero values. If we use Gaussian random variables as our sensing matrix A, we will not lose the data. It will not look the same for now, but it is still encoded in the output. In equation (8.1.5), the sampled points obtain only 3 output values in a different manner random measurements.

$$
\begin{bmatrix} 0.2336 \\ -0.1767 \\ -0.1572 \end{bmatrix} = \begin{bmatrix} 0.52 & 0.67 & 0.65 & 0.53 & -2.30 & -0.89 & 0.85 & -0.11 \\ 0.21 & -0.84 & -1.62 & -2.06 & -1.16 & -0.02 & 0.21 & 0.93 \\ -1.34 & -1.38 & -0.37 & -0.22 & 2.09 & -0.62 & 0.93 & -1.40 \end{bmatrix} \begin{bmatrix} 0 \\ 0.18 \\ 0.03 \\ 0 \\ 0 \\ 0 \\ 0.11 \\ 0 \end{bmatrix}
$$

$$(8.1.5)$$

Simple simulation examples：

1. A sparse time-domain signal(see Figure 8.1.3)

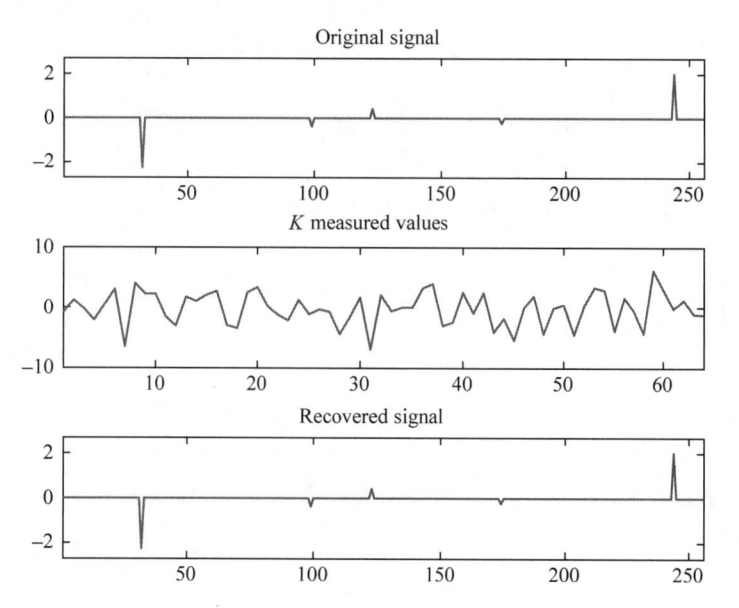

Figure 8.1.3　The simulation of a sparse time-domain signal with the compressive sensing method

The original signal had 256 values but we recovered it perfectly even though we only took 64 random measurements. As can be seen, compressed sensing worked for this simple example.

2. The image processing with CS method

Take the picture Lena as an example：the compressive sensing method is used for this image，here the OMP algorithm is adopted to recovery picture，as shown in Figure 8.1.4.

8.1.3　Application

CS is being a growing field and a wide variety of applications has benefited from this sensing modality. Figure 8.1.5 shows a taxonomy（分类）listing major applications of CS. This section overviews the application areas where CS finds its applicability in current scenario. This may be helpful in identifying an application area to work on using CS[4].

Original image　　　　　　　　　　　Recovery image

Figure 8. 1. 4　The original image and the recovery image based on CS method

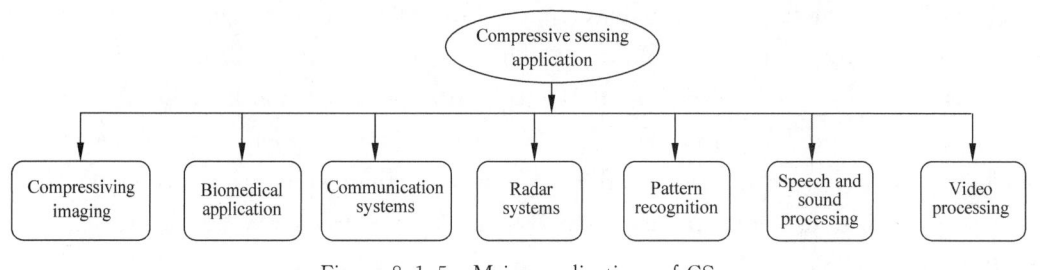

Figure 8. 1. 5　Major applications of CS

The various types of radar imaging techniques where CS has been used are synthetic aperture radar (SAR), inverse synthetic aperture radar (ISAR), through the wall imaging radar (TWR) and ground penetrating radar imaging (GPR).

In communication system, the research is mainly focused on the following aspects:

1. Wireless sensor networks(WSN)

The efficient data gathering schemes based on CS has been proposed for WSN in exploiting raw data compressibility using opportunistic routing. These compressive data gathering schemes offers advantages like robustness, prolonged network lifetime, reduced energy consumption and simple routing scheme, etc.

2. UWB communication

UWB communication basically makes use of CS architecture called RMPI, for acquisition of UWB signals. The reconstruction of original signal can be done by exploiting its spatial and temporal information. The other issues like, impulse radio detection, echo detection, channel estimation, high precision ranging and non-coherent UWB systems has also been addressed using CS.

3. Direction of arrival (DOA) estimation

Method for compressive beamforming using random projections of the sensor data for DOA estimation has been proposed. CS also has been used to solve problems in beamforming like grid-mismatch, reducing the number of sensors, DOA estimation for non-circular sources and the arrays with multiple co-prime frequencies.

4. Blind source separation (BSS)

CS for BSS addresses the separation of signal sources from the mixed music/speech signal using two-stage cluster-then-ℓ_1 optimization approach and using non-negative matrix factorization.

8.2   Ultra wideband（UWB）system

8.2.1   The definition of UWB

Ultra wideband （UWB） communication is based on the transmission of very short pulses with relatively low energy. This technology may see increased use in the field of wireless communications and ranging in the near future. UWB technique has a fine time resolution which makes it a technology appropriate for accurate ranging. UWB is usually used in short-range wireless applications but can be sent over wires. Ultra wideband advantages are that it can carry high data rates with low power and little interference.

Because of the huge bandwidth，UWB waves have a good material penetration capability. UWB is a communications technology that employs a wide bandwidth （typically defined as greater than 20% of the centre frequency or 500MHz）. The mathematical definition of relative bandwidth is[5]：

$$B_f = \frac{f_H - f_L}{(f_H + f_L)/2} = \frac{f_H - f_L}{f_c} > 20\% \tag{8.2.1}$$

The definition of absolute bandwidth is

$$B_f = f_H - f_L > 500\text{MHz} \tag{8.2.2}$$

According to Shannon's capacity formula，this large bandwidth provides a very high capacity. Thus，high processing gains can be achieved that allow the access of a large number of users to the system. Figure 8.2.1 shows the comparison between UWB and narrowband signal. As can be seen,the widthband of UWB system is really far wider than the narrowband system.

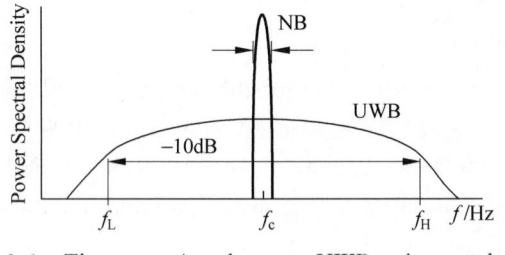

Figure 8.2.1   The comparison between UWB and narrowband signal

The impulse radio UWB is a carrier-less （i.e.，baseband） radio technology and accordingly，in this radio technique no mixer is needed. Therefore，the implementation of such a system is simple，which means that low cost transmitters/receivers can be achieved when compared to the conventional radio frequency （RF） carrier systems. Through the years （1960s~1990s） the United States military developed the UWB technology that was first used for ground penetrating radar. In 1998，the Federal Communication Commissions

(FCC) recognized the significance of UWB technology and initiated the regulatory review process of the technology. Consequently, in February 2002 the FCC report appeared, in which UWB technology was authorized for the commercial uses with different applications, operating frequency bands as well as the transmitted power spectral densities.

8.2.2　Comparison with other wireless communications

UWB, which stands for ultra-wideband, is a network standard that specifies how two UWB devices use short-range radio waves to communicate at high speeds with each other. For optimal communications, the devices should be within 2 to 10 meters (about 6.5 to 33 feet) of each other. Examples of UWB applications include wirelessly transferring video from a digital video camera, printing pictures from a digital camera, downloading media to a portable media player, or displaying a slide show on a projector. Figure 8.2.2 shows the spectrum of different wireless communication.

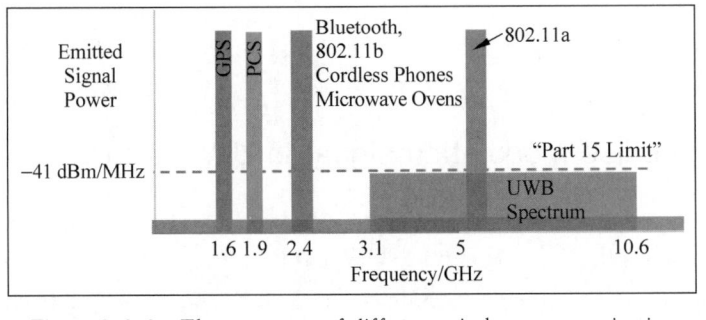

Figure 8.2.2　The spectrum of different wireless communication

Bluetooth is a standard, specifics protocol, that defines how two Bluetooth devices use short-range radio waves to transmit data. To communicate with each other, Bluetooth devices often must be within about 10 meters (about 33 feet) but can be extended to 100 meters with additional equipment. Examples of Bluetooth devices can include desktop computers, notebook computers, handheld computers, smart phones, headsets, microphones, digital cameras, GPS receivers, and printers.

Difference between UWB and Bluetooth is that UWB, which stands for ultra-wideband, is a network standard that specifies how two UWB devices use short-range radio waves to communicate at high speeds with each other. For optimal communications, the devices should be within 2 to 10 meters of each other.

8.2.3　Feature of UWB

Following are the **features** of UWB technology:
- Coverage range: About 30 meters;
- Data rate: About 1Gb/s;
- Operating frequency: below 1GHz, 3~5GHz and 6~10GHz;

- Standard: IEEE 802. 15. 4a;
- PHYSICAL Layer: MB OFDM, DS-UWB;
- Modulation types: BPM, OOK, PAM, OPM.

Following are the **advantages** of UWB:

- Low power;
- Good noise immunity;
- Signals can penetrate variety of materials easily;
- High immunity to multipath fading;
- Potentially very high data rates.

Disadvantages of UWB:

- Higher cost;
- Slower adoption rate;
- Long signal acquisition times;
- FCC has limited emission requirements which is less than 0. 5mW max power over 7. 5GHz band;
- The UWB technology has issues of co-existence and interference with other radio based technologies.

8.2.4 Impulse modulation signal in UWB signal

The selection of impulse-signal types for UWB impulse systems is one of the fundamental considerations in designing UWB impulse systems, antennas, and circuits because the type of an impulse determines the UWB signal's spectrum characteristic. Many types of impulse signals such as step pulse, Gaussian-like (or monopolar) impulse, Gaussian-like single-cycle (or monocycle) pulse, Gaussian-like doublet pulse, and multi-cycle pulse can be used for UWB impulse systems. Among those, Gaussian-like impulse, doublet pulse, and monocycle pulse (see Figure 8. 2. 3) are typically used in UWB impulse systems. Particularly, the monocycle pulse is preferred in most UWB impulse systems because of its spectral characteristics (having no DC level) that facilitate easier wireless transmission than the impulse.

UWB uses various modulation schemes based on application requirements. There are time based techniques and shape based techniques. Pulse position modulation (PPM) is used as time based modulation. PPM is a simple technique but it needs fine time resolution in receiver. BPM (bi-phase modulation), OOK (on-off keying), PAM (pulse amplitude modulation), OPM (orthogonal pulse modulation) etc. are used as shaped based modulation techniques. PAM and OOK are also simple binary only techniques but they have poor noise immunity. OPM is complex but it has the advantage of orthogonality.

According to the principle of different modulations, try to match the following modulations with each corresponding waveform(连线匹配).

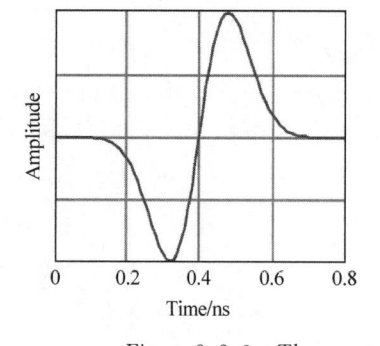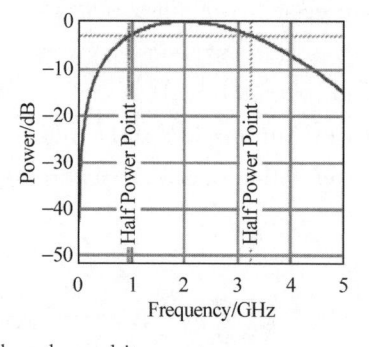

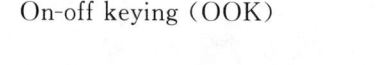

Figure 8.2.3　The monocycle pulse and its spectrum

On-off keying (OOK)

Pulse amplitude modulation (PAM)

Bi-phase modulation (BPM)

Pulse position modulation (PPM)

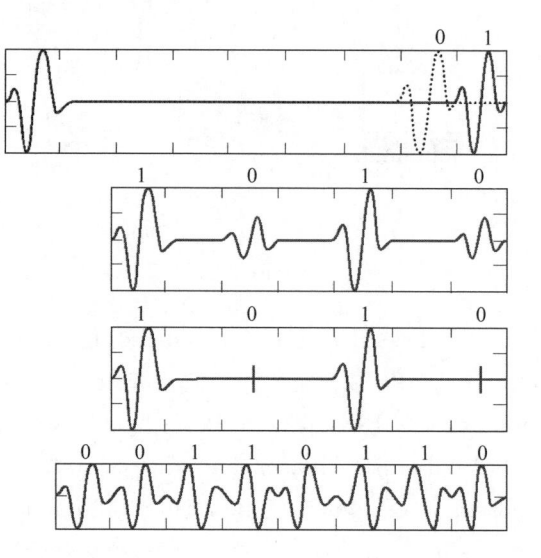

8.2.5　The application of UWB

Ultra-wideband characteristics are well-suited to short-distance applications, such as PC peripherals. Due to low emission levels permitted by regulatory agencies, UWB systems tend to be short-range indoor applications. Due to the short duration of UWB pulses, it is easier to design high data rates; data rate may be exchanged for range by aggregating pulse energy per data bit (with integration or coding techniques). Conventional orthogonal frequency-division multiplexing (OFDM) technology may also be used, subject to minimum-bandwidth requirements. High-data-rate UWB may enable wireless monitors, the efficient transfer of data from digital camcorders, wireless printing of digital pictures from a camera without the need for a personal computer and file transfers between cell-phone handsets and handheld devices such as portable media players. UWB is used for real-time location systems; its precision capabilities and low power make it well-suited for radio-frequency-sensitive environments, such as hospitals. Another feature of UWB is its short broadcast time.

Ultra-wideband is also used in "see-through-the-wall" precision radar-imaging technology, precision locating and tracking (using distance measurements between radios), and precision time-of-arrival-based localization approaches. The system of UWB through-the-wall radar system is given in Figure 8. 2. 4 It is efficient, with a spatial capacity of approximately $1013b/s/m^2$. UWB radar has been proposed as the active sensor component in an Automatic Target Recognition application, designed to detect humans or objects that have fallen onto subway tracks[6].

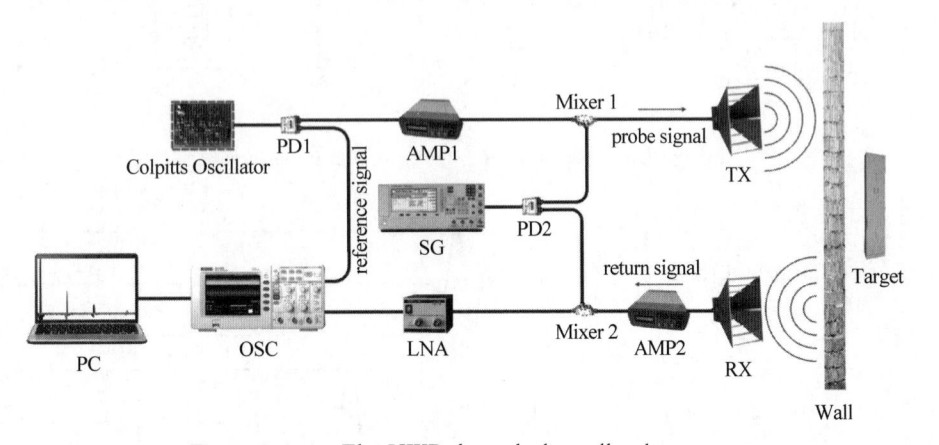

Figure 8. 2. 4　The UWB through-the-wall radar system

UWB has also been considered for development of personal area networks and appeared in the IEEE 802. 15. 3a draft PAN standard. However, after several years of deadlock, the IEEE 802. 15. 3a task group was dissolved in 2006. The work was completed by the WiMedia Alliance and the USB Implementer Forum. Slow progress in UWB standards development, the cost of initial implementation, and significant lower than expected performance are several reasons for the limited use of UWB in consumer products (which caused several UWB vendors to cease operations in 2008 and 2009).

8.3　MIMO technology

MIMO (multiple input multiple output) is a technique where multiple antennas are used at both the transmitter and the receiver to increase the link reliability and the spectral efficiency. This concept has been around for many years but its application in wireless standards is more recent. This is probably due in part to the fact that OFDM (orthogonal frequency-division multiplexing), which facilitates the implementation of MIMO, is now commonly used in today's wireless standards. MIMO techniques are used in technologies like Wi-Fi and LTE, and new techniques are under study for future standards like LTE Advanced[7].

By transmitting the same data on multiple streams, the MIMO radios introduce redundancy into data transmission that classic single antenna setups (SISO: single in, single out) can not provide. This gives MIMO systems several advantages over typical SISO configurations:

(1) MIMO radios can utilize the bounced and reflected RF transmission to actually improve the signal strength even without clear line-of-sight, since MIMO radios receive and combine multiple streams of the same data. This is particularly useful in urban environments, where signal degradation between single antennas without clear line-of-sight is a major issue.

(2) Overall throughput can be improved, allowing for greater quality and quantity of video or other data to be sent over the network.

(3) By utilizing multiple data streams, lost or dropped data packets can be reduced, resulting in better video or audio quality.

8.3.1   The model of MIMO

The basic MIMO model is in Figure 8.3.1. Suppose there are M transmitter antennas and N receiver antennas.

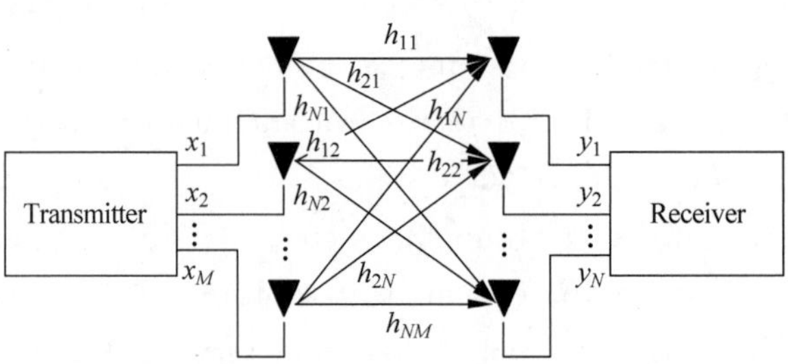

Figure 8.3.1   The basic model of MIMO system

The received signal can be written as:

$$y = Hx + n \tag{8.3.1}$$

where the transmitting signal is $x = [x_1, x_2, \cdots, x_M]$, the receiving signal is $y = [y_1, y_2, \cdots, y_N]$, the channel is $H = \begin{bmatrix} h_{11} & h_{12} & \cdots & h_{1M} \\ \vdots & \vdots & & \vdots \\ h_{N1} & h_{N2} & \cdots & h_{NM} \end{bmatrix}$ and n is the noise in the system.

Data to be transmitted is divided into independent data streams. The number of streams M is always less than or equal to the number of antennas N; in the case of asymmetrical ($m \neq n$) antenna constellations, it is always smaller or equal the minimum number of antennas. For example, a 4×4 system could be used to transmit four or fewer streams, while a 3×2 system could transmit two or fewer streams. When M antennas are used at both the transmitter and the receiver, and M is large, then the theoretical maximum capacity C increases linearly with the number of streams M[8].

$$C = MB \log_2 \left(1 + \frac{S}{N}\right) \tag{8.3.2}$$

According to the number of users, there are two types systems: SU-MIMO and MU-MIMO, as shown in Figure 8.3.2.

(1) When the data rate is to be increased for a single UE, this is called Single User

MIMO（SU-MIMO）；

（2）When the individual streams are assigned to various users，this is called multi user MIMO（MU-MIMO）. This mode is particularly useful in the uplink because the complexity on the UE side can be kept to a minimum by using only one transmit antenna. This is also called "collaborative MIMO".

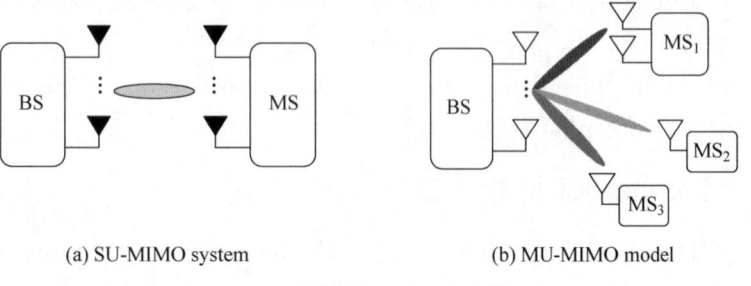

(a) SU-MIMO system   (b) MU-MIMO model

Figure 8.3.2   SU-MIMO and MU-MIMO systems

Antenna technologies are the key in increasing network capacity. Beamforming is the method used to create the radiation pattern of an antenna array. It can be applied in all antenna array systems as well as MIMO systems.

In MIMO systems，the space-time block-codes（STBC）are widely used，which involves the transmission of multiple redundant copies of data to compensate for fading and thermal noise in the hope that some of them may arrive at the receiver in a better state than others. In the case of STBC in particular，the data stream to be transmitted is encoded in blocks，which are distributed among spaced antennas and across time. An STBC is usually represented by a matrix. Each row represents a time slot and each column represents one antenna's transmissions over time. Space-time codes combine spatial and temporal signal copies as illustrated in Figure 8.3.3. The signals s_1 and s_2 are multiplexed in two data chains. After that，a signal replication is added to create the Alamouti space-time block code.

$$\text{Time-slots} \begin{bmatrix} s_{11} & s_{12} & \cdots & s_{1n_T} \\ s_{21} & s_{22} & \cdots & s_{2n_T} \\ \vdots & \vdots & & \vdots \\ s_{T1} & s_{T2} & \cdots & s_{Tn_T} \end{bmatrix} \qquad S = \begin{bmatrix} S_1 & S_2 \\ -S_2^* & S_1^* \end{bmatrix}$$

Transmit antennas

(a) Space-time character   (b) STBC for 2 antennas

Figure 8.3.3   The space-time codes

8.3.2   Applications

1. LTE

LTE can use transmit diversity（MISO）and receive diversity（SIMO）as well as beamforming, either alone or in combination with MIMO. For LTE and LTE Advanced, successive 3GPP standards releases increase transmission complexity, beginning with release 8, which introduced transmission mode 7 (TM7), which supports single layer beamforming. Release 9 added TM8, which supports dual layer beamforming (i. e. 2×2 MIMO with beamforming) and Release 10 adds TM9, which supports up to 8×8 MIMO with beamforming. The conceptual diagram of adopted 8×8 LTE terms is illustrated in Figure 8. 3. 4.

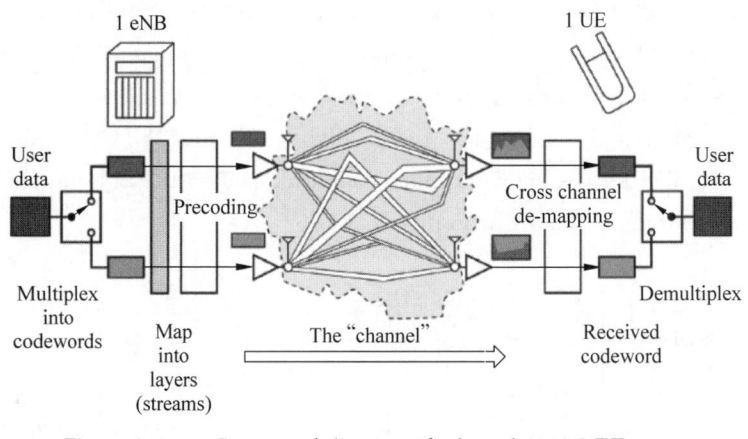

Figure 8. 3. 4　Conceptual diagram of adopted 8×8 LTE terms

The terms "codeword" "layer" "precoding" and "beamforming" have each been adopted specifically for LTE to refer to signals and their processing. The terms shown in Figure 8. 3. 4 are used in the following ways:
- **Codeword**: A codeword represents user data before it is formatted for transmission. In the most common case of single-user MIMO (SU-MIMO), up to two codewords are sent to a single handset or user equipment (UE).
- **Layer** (or stream): For MIMO, at least two layers must be used. Up to four for LTE and eight for LTE-A are allowed. The number of layers is always less than or equal to the number of antenna ports. The receiver needs at least as many antenna ports as the number of layers.
- **Precoding**: Precoding modifies the layer signals before transmission. This may be done for diversity, beamforming, or spatial multiplexing. The MIMO channel conditions may favor one layer (data stream) over another. If the spatial multiplexing is closed loop, the UE provides a precoding matrix indicator (PMI) so the eNB can cross-couple the streams to counteract the imbalance in the channel.
- **Beamforming**: Beamforming modifies the transmit signals to give the best carrier-to-noise interference plus noise ratio (CINR) at the output of the channel, normally by maximizing antenna gain in the direction of a particular UE. It may also be set to minimize gain in the direction of a second UE which is managed by another base station (eNB).

The basic concept for LTE in downlink is OFDMA (Uplink: SC-FDMA), while MIMO technologies are an integral part of LTE. Modulation modes are QPSK, 16QAM, and 64QAM. Peak data rates of up to 300 Mb/s (4×4 MIMO) and up to 150 Mb/s (2×2 MIMO) in the downlink and up to 75 Mb/s in the uplink are specified.

In order to keep the complexity low at the UE end, MU-MIMO is used in the uplink. To do this, multiple UEs, each with only one Tx antenna, use the same channel.

2. WiMAX

The WiMAX 802. 16e-2005 standard specifies MIMO in Wireless MAN-OFDMA mode. This standard defines a large number of different matrices for coding and distributing to antennas. In principle, two, three or four TX antennas are possible. For all modes, the matrices A, B, and C are available. In the "STC encoder" block, the streams are multiplied by the selected matrix and mapped to the antennas. The downlink MIMO WiMAX system is shown in Figure 8. 3. 5.

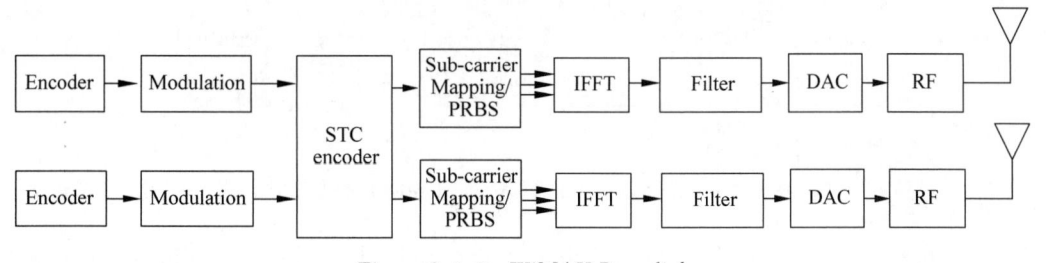

Figure 8. 3. 5　WiMAX Downlink

In Uplink-MIMO only different pilot patterns are used. Coding and mapping is the same like in non-MIMO case. In addition to single user MIMO (SU-MIMO) two different user can use the same channel (collaborative MIMO, MU-MIMO).

3. Radar

Like MIMO communications, MIMO radar offers a new paradigm for signal processing research. MIMO radar possesses significant potentials for fading mitigation, resolution enhancement, and interference and jamming suppression. Fully exploiting these potentials can result in much improved target detection and recognition performance.

Known MIMO radars may be divided into two classes, as shown in Figure 8. 3. 6:

(1) MIMO radars with collocated antennas and coded signals;

(2) Radars with widely separated antennas, the so-called "Statistical MIMO radars".

The main areas of research are focused on:

(1) Synthetic aperture radar (SAR) MIMO;

(2) MIMO radar using compressive sampling;

(3) UWB MIMO radar for through the wall and medical imaging;

(4) Performance of MIMO radar with angular estimation and targets tracking;

(5) Signaling strategies for hybrid MIMO phased-array radar.

One of the important applications of MIMO SAR is on high resolution microwave imaging,

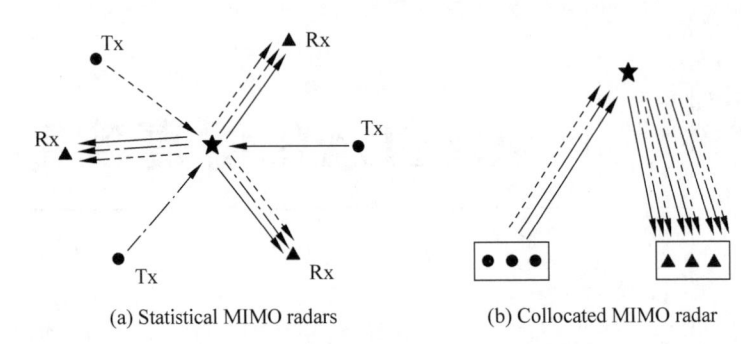

(a) Statistical MIMO radars　　　　(b) Collocated MIMO radar

Figure 8.3.6　MIMO radar models

a technology which can connect to a huge variety of emerging industries including airport security checks as well as auto piloting. Following Figure 8.3.7 is the 3D imaging results[9].

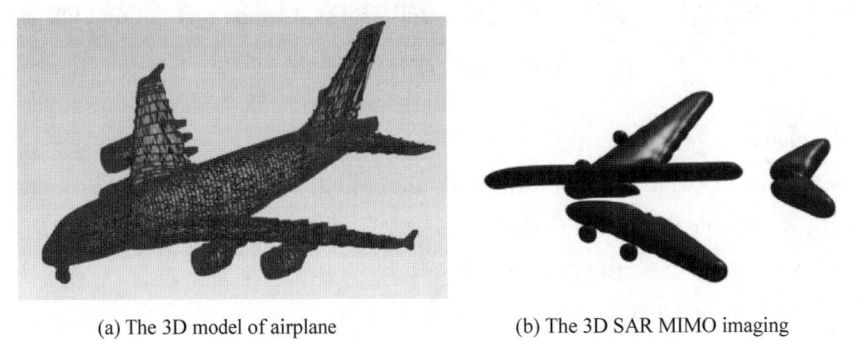

(a) The 3D model of airplane　　　　(b) The 3D SAR MIMO imaging

Figure 8.3.7　The comparison of 3D imaging

附录 A	MATLAB 仿真实验

Experiment 1　Continuous-wave modulation（corresponding to Chapter 4）

1. Introduction

The principles of the CW modulation, including the AM/DSB/FM communication systems will be studied in this experiment.

At the end of this experiment, you should have learned:

- the simple procedure of CW modulation.
- the concepts of modulation and demodulation.
- the effects of proportional, integral and derivative actions.
- how to carry out simulations using MATLAB software.

2. Description of CW modulation

1）Block diagram

The CW modulation system is illustrated in Figure A. 1.

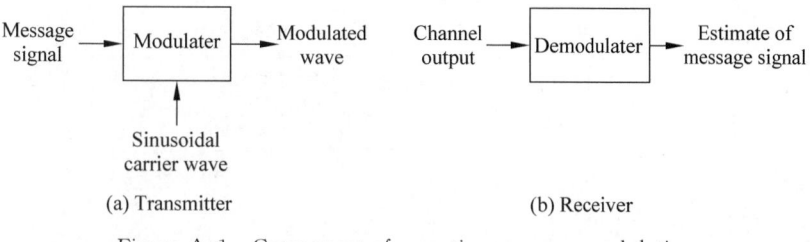

(a) Transmitter　　　　　　　　　　　　(b) Receiver

Figure A. 1　Component of a continuous-wave modulation

In basic signal processing terms, we find that the transmitter of an analog communication system consists of a modulator and the receiver consists of a demodulator, as depicted in Figure A. 1. In additional to the signal received from the transmitter, the receiver input includes channel noise.

For the demodulation, here we use the coherent detection to recover the message signal $m(t)$, as can be seen in Figure A. 2.

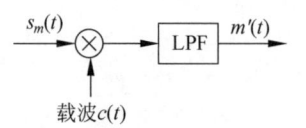

Figure A. 2   The demodulation of coherent detection

2）The model of AM system and simulation

The generation principle of AM is shown in Figure A. 3.

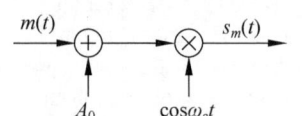

Figure A. 3   The generation of AM

Suppose the modulating signal $m(t)$ is single tone signal $\cos\omega_c t$, so the modulated signal can be given by：

$$S_{AM}(t) = A_0 \cos\omega_c t + m(t)\cos\omega_c t \tag{1}$$

3）The model of DSB system （see Figure A. 4）

Figure A. 4   The modulation of DSB

4）The principle of FM system and simulation

The frequency modulated signal is described in the time domain by

$$s(t) = A_c \cos\left[2\pi f_c t + 2\pi k_f \int_0^t m(\tau)\mathrm{d}\tau\right] \tag{2}$$

The time domain and frequency domain of an FM signal are plotted in Figure A. 5.

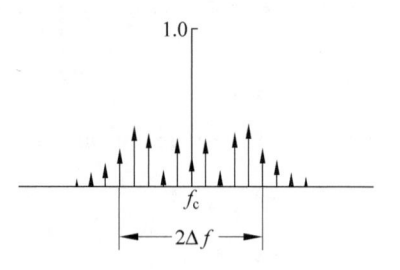

(a)Time domain   (b) Frequency domain

Figure A. 5   Time domain and frequency domains waveforms of a FM signal

3. MATLAB simulations for CW modulation

1）AM simulation

Step 1：start MATLAB and type the following commands at the prompt

```
>> a0 = 2; f0 = 10; fc = 50; fs = 1000; snr = 5;
t = [ - 20:0.001:20];
am1 = cos(2 * pi * f0 * t);          % message signal
am = a0 + am1;
t1 = cos(2 * pi * fc * t);           % carrier wave
s_am = am. * t1;
AM1 = fft(am1); T1 = fft(t1); S_AM = fft(s_am);
f = (0:40000) * fs/40001 - fs/2;
subplot(3,2,1); plot(t(19801:20200),am1(19801:20200)); title('message signal');
subplot(3,2,2); plot(f,fftshift(abs(AM1))); title('spectrum of message signal ');
subplot(3,2,3); plot(t(19801:20200),t1(19801:20200)); title('carrier wave ');
subplot(3,2,4); plot(f,fftshift(abs(T1))); title(' spectrum of carrier wave ');
subplot(3,2,5); plot(t(19801:20200),s_am(19801:20200)); title('modulated signal');
subplot(3,2,6); plot(f,fftshift(abs(S_AM))); title('spectrum of modulated signal');
```

Click "Enter"，you can get the simulations of the AM process. Copy the results in the following blank（see Figure A. 6）.

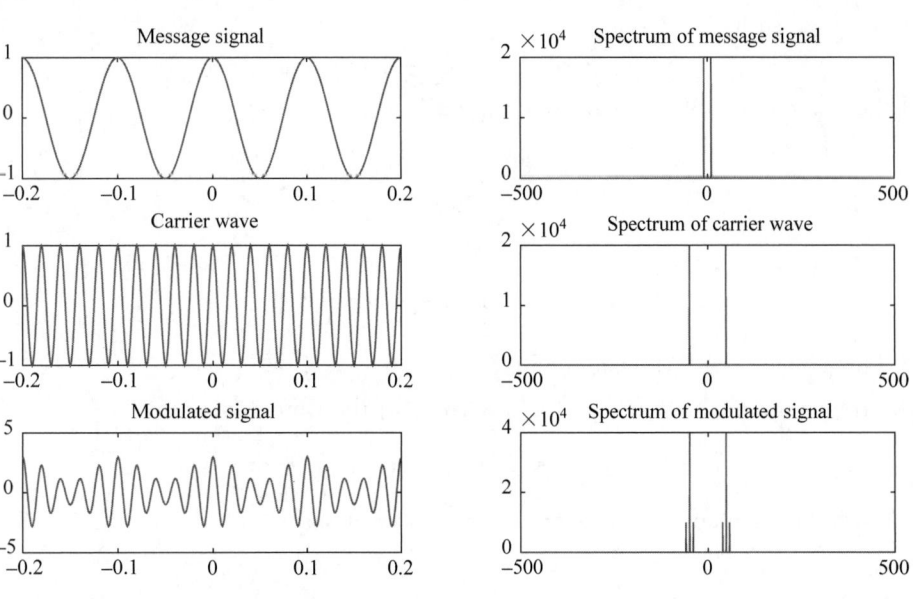

Figure A. 6　AM simulation results

Step 2：add the noise to the modulated signal

```
>> % Add noise
y = awgn(s_am,snr);  % 高斯白噪声
a = [35,65]; b = [30,70];
Wp = a/(fs/2); Ws = b/(fs/2); Rp = 3; Rs = 15;
[N,Wn] = Buttord(Wp,Ws,Rp,Rs) ;
[B,A] = Butter(N,Wn,'bandpass');
q = filtfilt(B,A,y);
Q = fft(q); Y = fft(y);
subplot(2,2,1); plot(t(19801:20200),y(19801:20200)); title('signal with addictive noise ');
subplot(2,2,2); plot(f,fftshift(abs(Y))); title(' signal spectrum with addictive noise ');
subplot(2,2,3); plot(t(19801:20200),q(19801:20200)); title('The signal after the BPF ');
```

```
subplot(2,2,4); plot(f,fftshift(abs(Q)));title('The signal spectrum after the BPF ');
```

Click "Enter"，you can get the simulations of the AM signal with noise. Copy the results in the following blank（see Figure A. 7）.

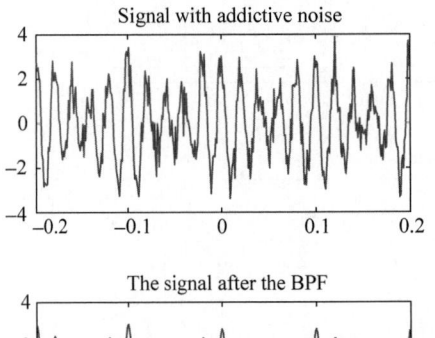
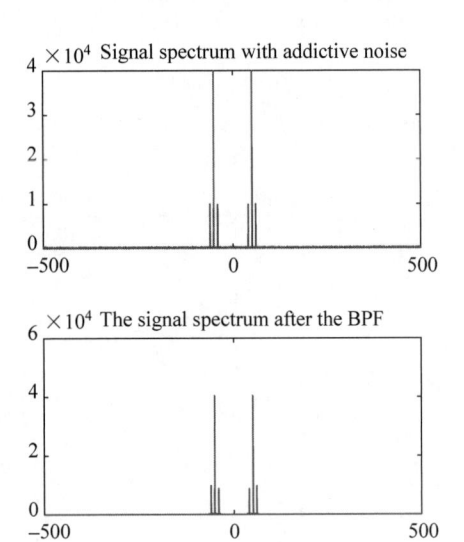
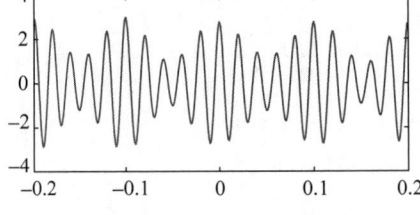

Figure A. 7　The modulated signal with noise

Step 3：the demodulation process

```
>> ss_am = q. * t1;
Wp = 15/(fs/2);Ws = 40/(fs/2);Rp = 3; Rs = 20;
[N,Wn] = Buttord(Wp,Ws,Rp,Rs) ;
[B,A] = Butter(N,Wn,'low');
m0 = filtfilt(B,A,ss_am);
M0 = fft(m0);
subplot(2,1,1);plot(t(19801:20200),m0(19801:20200));title('Demodulated signal');
subplot(2,1,2); plot(f,fftshift(abs(M0)));title('spectrum of demodulated signal');
```

The demodulated signal result of AM modulation signal can be get. Copy the result in the following blank（see Figure A. 8）.

2）DSB simulation

Step 1：the signal generation at the transmitter

```
>> dt = 0.0000001;
t = 0:dt:0.005;
f1 = 2000; % message frequency
m = cos(2 * pi * f1 * t);
figure(1);
subplot(311);plot(t,m);title('模拟信源(a)');axis([0,0.001, - 1.5,1.5]);
%%%%%% carrier wave          %%%%%%%%%%%%%%
fc = 1000000; % carrier frequency
c = cos(2 * pi * fc * t);
subplot(312);plot(t,c);title('载波信号(b)');axis([0,0.00001, - 1.5,1.5]);
%%%%%%%%%%%% DSB modulation %%%%%%%%%%%%%
```

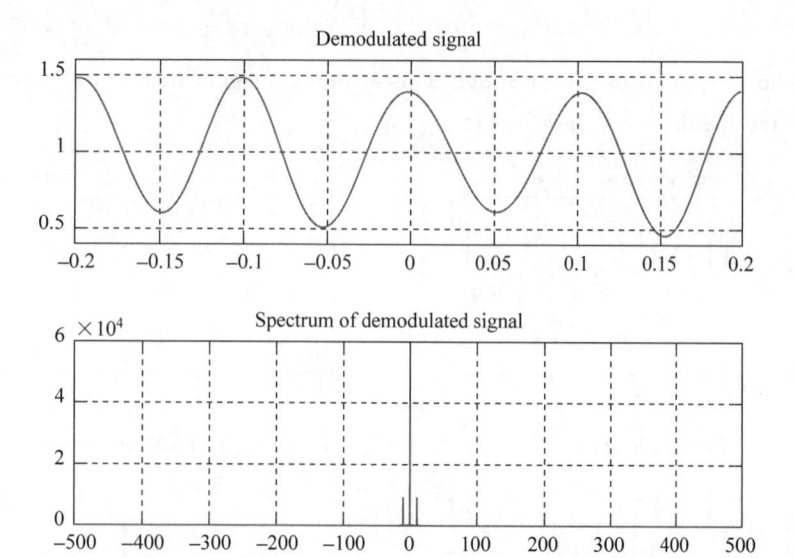

Figure A. 8　The recovery signal and its spectrum

```
s = m. * c; % DSB signal
subplot(313);plot(t,s);title('DSB 信号(c)');axis([0,0.001, - 1.5,1.5]);
```

You can get the simulation of the DSB signal. Copy the result in the following blank（see Figure A. 9）.

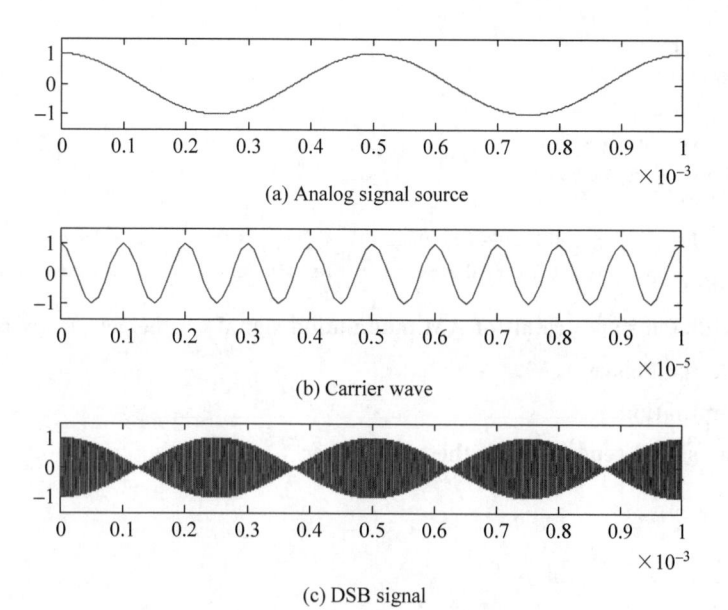

Figure A. 9　The simulation of DSB signal

Step 2：the multipath propagation and noise in channel

```
>> a1 = 0.9; % 直射波衰减
a2 = 0.7; % 反射波衰减
t2 = 0.0001; % 反射波时延
```

```
s1 = a1 * s; % 接收直射波
s2 = a2 * cos(2 * pi * f1 * (t + t2)). * cos(2 * pi * fc * (t + t2)); % 接收反射波
x = s1 + s2; % 总接收信号
x1 = awgn(x,10); % 叠加高斯白噪声,信噪比为 10
figure(2);
plot(t,x1);title('接收端信号(多径 + 噪声)(f)');axis([0,0.001, - 3,3]);
```

You can get the simulation of the AM signal with multipath & noise. Copy the result in the following blank(see Figure A. 10).

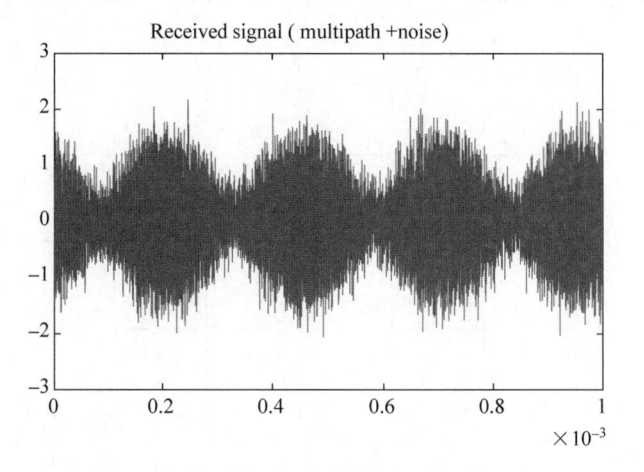

Figure A. 10　The multipath signal with noise

Step 3: the band pass filter

```
>> w1 = 2 * dt * (fc - f1 - 20);
w2 = 2 * dt * (fc + f1 + 20);
[c,d] = butter(4,[w1 w2],'bandpass'); % 4 阶 Butterworth 滤波器
x2 = filter(c,d,x1); % 信号通过带通滤波器
figure(3);
plot(t,x2);title('信号通过带通滤波器(d)');axis([0,0.003, - 1.5,1.5]);
```

Copy the result in the following blank(see Figure A. 11).

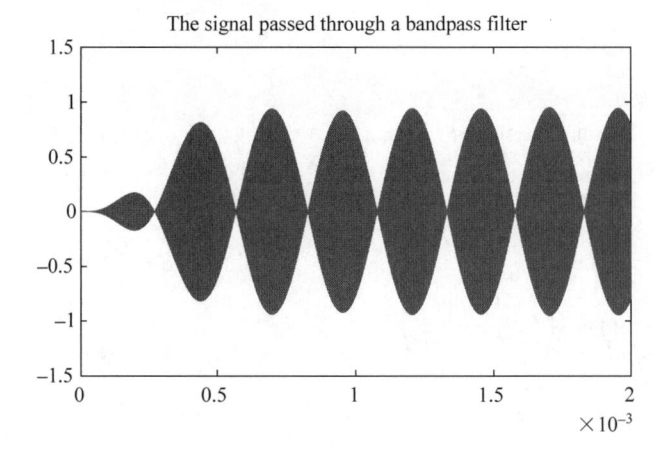

Figure A. 11　The output signal of band pass

Step 4：demodulation

```
>> x3 = cos(2 * pi * fc * t). * x2; % 与载波相乘
w = 2 * dt * (f1 + 500);
[p,q] = butter(4,w,'low'); % 4 阶 Butterworth 低通滤波器
x4 = filter(p,q,x3); % 信号通过低通滤波器
figure(4);
plot(t,x4);title('信号相干解调、通过低通滤波器 ');axis([0,0.003, - 1,1]);
```

Copy the final recovery signal of the DSB system in the blank（see Figure A. 12）.

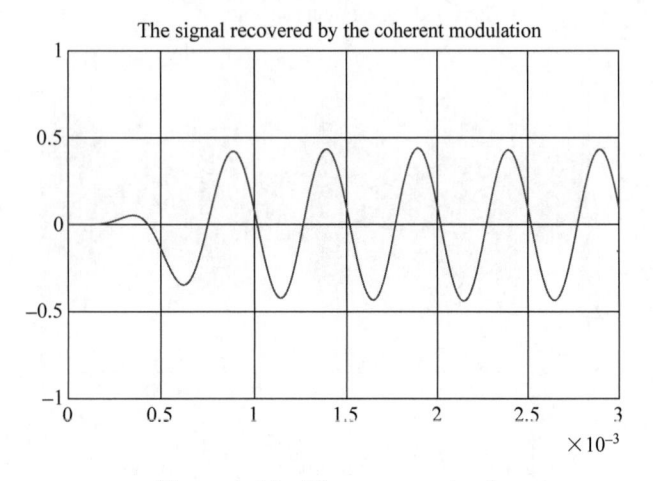

Figure A. 12　The recovery signal

3）FM simulation

Step 1：the generation of FM signal（time domain）

```
>> dt = 0.001;                                    % 设定时间步长
t = 0:dt:1.5;                                      % 产生时间向量
am = 5;                                            % 设定调制信号幅度
fm = 5;                                            % 设定调制信号频率
mt = am * cos(2 * pi * fm * t);                    % 生成调制信号
fc = 50;                                           % 设定载波频率
ct = cos(2 * pi * fc * t);                         % 生成载波
kf = 10;                                           % 设定调频指数
int_mt(1) = 0;
for i = 1:length(t) - 1
    int_mt(i + 1) = int_mt(i) + mt(i) * dt;        % 求信号 m(t)的积分
end                                                % 调制,产生已调信号
sfm = am * cos(2 * pi * fc * t + 2 * pi * kf * int_mt); % 调制信号
figure(1)
subplot(3,1,1);plot(t,mt);                         % 绘制调制信号的时域图
xlabel('时间 t');
title('调制信号的时域图');
subplot(3,1,2);plot(t,ct);                         % 绘制载波的时域图
xlabel('时间 t');
title('载波的时域图');
subplot(3,1,3);
```

```
plot(t,sfm);                              % 绘制已调信号的时域图
xlabel('时间 t');
title('已调信号的时域图');
```

Click Enter，The time domain of the message signal/carrier wave/FM can be get. Copy the results in the next blank(see Figure A. 13).

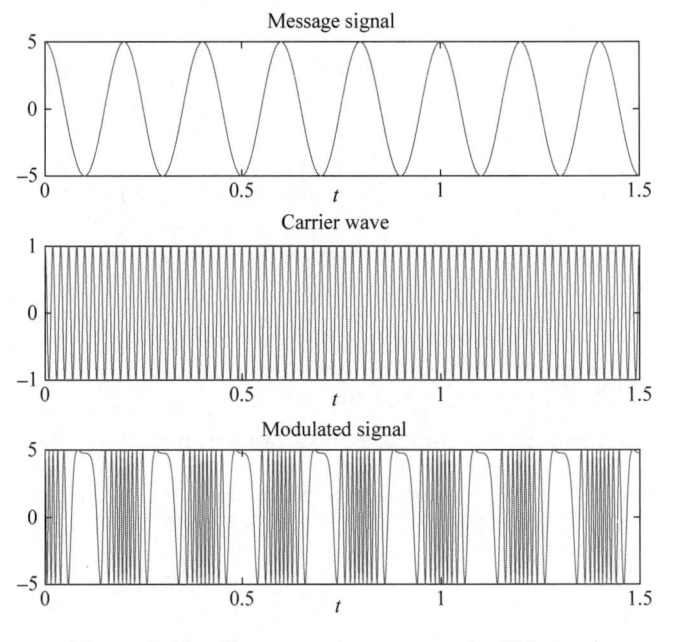

Figure A. 13   The process to generate the FM signal

Step 2：the FM signal（frequency domain of message signal and modulated signal）

```
>> ts = 0.001;                    % 抽样间隔
fs = 1/ts;                        % 抽样频率
df = 0.25;                        % 所需的频率分辨率,用在求傅里叶变换时,它表示 FFT 的最小频率间隔
% ***** 对调制信号 m(t)求傅里叶变换 *****
m = am * cos(2 * pi * fm * t);    % 原调信号
fs = 1/ts;
if margin == 2
    n1 = 0;
else
    n1 = fs/df;
end
n2 = length(m);
n = 2^(max(nextpow2(n1),nextpow2(n2)));
M = fft(m,n);
m = [m,zeros(1,n - n2)];
df1 = fs/n;                       % 以上程序是对调制后的信号求傅里叶变换
M = M/fs;                         % 缩放,便于在频谱图上整体观察
f = [0:df1:df1 * (length(m) - 1)] - fs/2;
fs = 1/ts;
if margin == 2
```

```
        n1 = 0;
    else
        n1 = fs/df;
    end
    n2 = length(sfm);
    n = 2^(max(nextpow2(n1),nextpow2(n2)));
    U = fft(sfm,n);
    u = [sfm,zeros(1,n-n2)];
    df1 = fs/n;                     % 以上是对已调信号求傅里叶变换
    U = U/fs;
    figure(2)
    subplot(2,1,1)
    plot(f,abs(fftshift(M)))        % fftshift:将 FFT 中的 DC 分量移到频谱中心
    xlabel('频率 f')
    title('原调制信号的频谱图')
    subplot(2,1,2)
    plot(f,abs(fftshift(U)))
    xlabel('频率 f')
    title('已调信号的频谱图')
```

Copy the spectrum results and compare these two spectrums(see Figure A. 14).

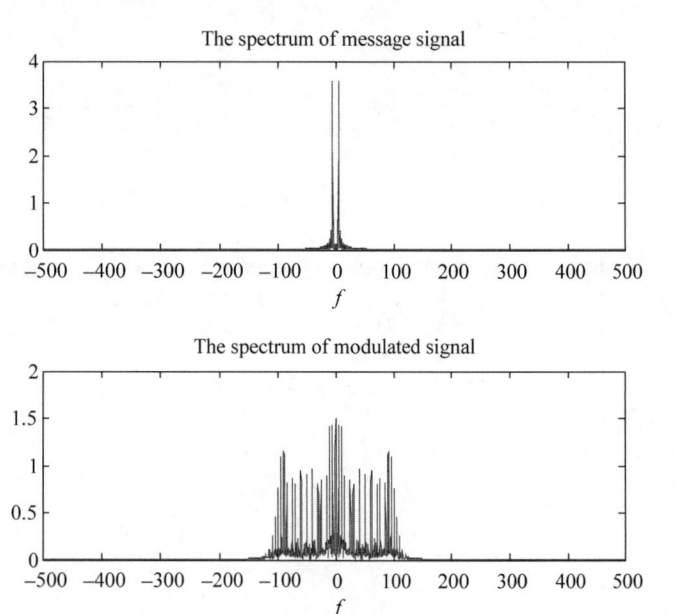

Figure A. 14   The spectral comparison before and after modulation

Step 3：add noise to FM signal（three types：low SNR/High SNR/no noise）

```
% ************* 添加高斯白噪声 *************
sn1 = 10;                          % 设定信噪比(小信噪比)
sn2 = 30;                          % 设定信噪比(大信噪比)
sn = 0;                            % 设定信噪比(无信噪比)
db = am^2/(2*(10^(sn/10)));        % 计算对应的高斯白噪声的方差
n = sqrt(db)*randn(size(t));       % 生成高斯白噪声
```

```
nsfm = n + sfm;                          % 生成含高斯白噪声的已调信号(信号通过信道传输)
```

Step 4：demodulation（without noise）

```
>> for i = 1:length(t) − 1          % 接收信号通过微分器处理
       diff_nsfm(i) = (nsfm(i + 1) − nsfm(i))./dt;
   end
diff_nsfmn = abs(hilbert(diff_nsfm));       % Hilbert 变换,求绝对值得到瞬时幅度(包络检波)
zero = (max(diff_nsfmn) − min(diff_nsfmn))/2;
diff_nsfmn1 = diff_nsfmn − zero;
figure(3)
subplot(3,1,1);plot(t,mt);          % 绘制调制信号的时域图
xlabel('时间 t');
title('调制信号的时域图');
subplot(3,1,2);plot(t,sfm);         % 绘制已调信号的时域图
xlabel('时间 t');
title('无噪声条件下已调信号的时域图');
nsfm = sfm;
for i = 1:length(t) − 1             % 接收信号通过微分器处理
    diff_nsfm(i) = (nsfm(i + 1) − nsfm(i))./dt;
end
diff_nsfmn = abs(hilbert(diff_nsfm));          % Hilbert 变换,求绝对值得到瞬时幅度(包络检波)
zero = (max(diff_nsfmn) − min(diff_nsfmn))/2;
diff_nsfmn1 = diff_nsfmn − zero;
subplot(3,1,3);                     % 绘制无噪声条件下解调信号的时域图
plot((1:length(diff_nsfmn1))./1000,diff_nsfmn1./400,'r');
xlabel('时间 t');
title('无噪声条件下解调信号的时域图');
```

Copy the simulation process of FM without noise in the following blank（see Figure A. 15）.

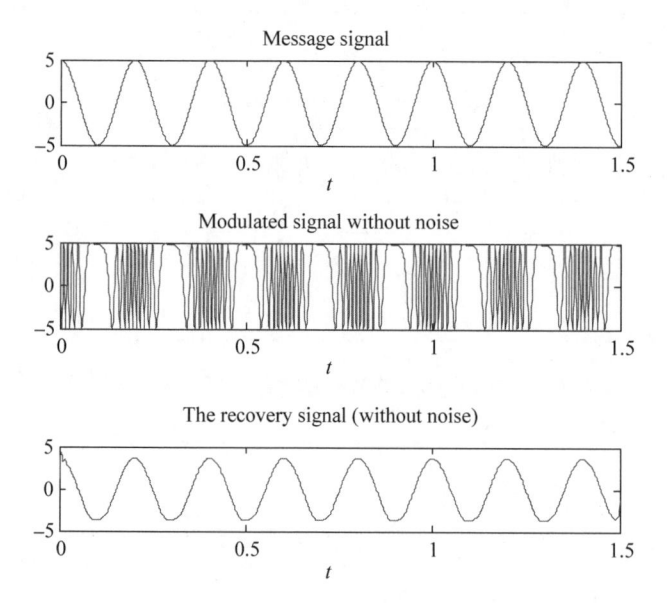

Figure A. 15　The recovery signal without noise

Step 5：demodulation（with low SNR）

```
>> figure(4)
subplot(3,1,1);plot(t,mt);          % 绘制调制信号的时域图
xlabel('时间 t');
title('调制信号的时域图');
db1 = am^2/(2 * (10^(sn1/10)));     % 计算对应的小信噪比高斯白噪声的方差
n1 = sqrt(db1) * randn(size(t));    % 生成高斯白噪声
nsfm1 = n1 + sfm;                   % 生成含高斯白噪声的已调信号(信号通过信道传输)
for i = 1:length(t) - 1             % 接收信号通过微分器处理
    diff_nsfm1(i) = (nsfm1(i + 1) - nsfm1(i))./dt;
end
diff_nsfmn1 = abs(hilbert(diff_nsfm1));     % Hilbert 变换,求绝对值得到瞬时幅度(包络检波)
zero = (max(diff_nsfmn) - min(diff_nsfmn))/2;
diff_nsfmn1 = diff_nsfmn1 - zero;
subplot(3,1,2);
plot(1:length(diff_nsfm),diff_nsfm);        % 绘制含小信噪比高斯白噪声已调信号的时域图
xlabel('时间 t');
title('含小信噪比高斯白噪声已调信号的时域图');
subplot(3,1,3);                     % 绘制含小信噪比高斯白噪声解调信号的时域图
plot((1:length(diff_nsfmn1))./1000,diff_nsfmn1./400,'r');
xlabel('时间 t');
title('含小信噪比高斯白噪声解调信号的时域图');
```

Copy the simulation process of FM with low SNR in the following blank（see Figure A. 16）.

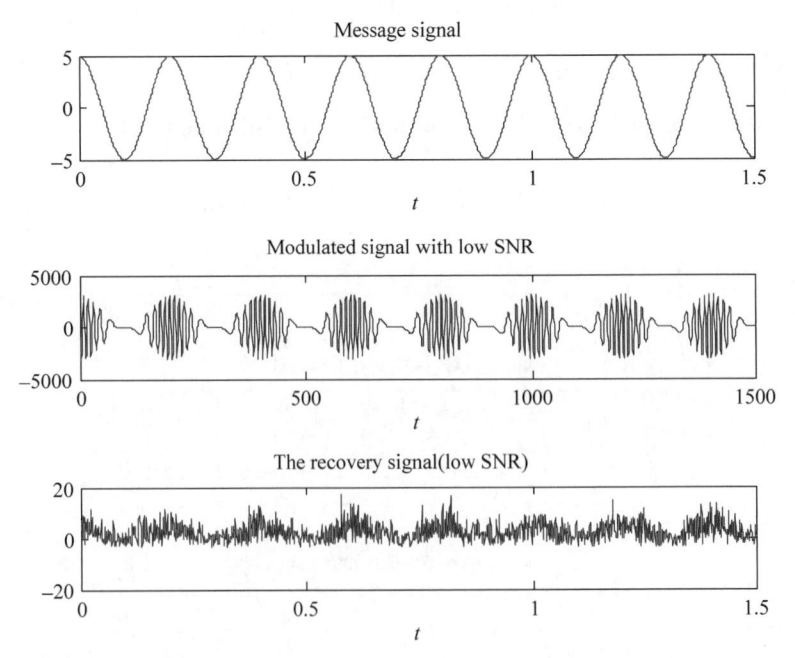

Figure A. 16　The recovery signal with low SNR

Step 6：demodulation（with high SNR）

```
>> figure(5)
```

```
subplot(3,1,1);plot(t,mt);              % 绘制调制信号的时域图
xlabel('时间 t');
title('调制信号的时域图');
db1 = am^2/(2 * (10^(sn2/10)));         % 计算对应的大信噪比高斯白噪声的方差
n1 = sqrt(db1) * randn(size(t));        % 生成高斯白噪声
nsfm1 = n1 + sfm;                       % 生成含高斯白噪声的已调信号(信号通过信道传输)
for i = 1:length(t) - 1                 % 接收信号通过微分器处理
    diff_nsfm1(i) = (nsfm1(i + 1) - nsfm1(i))./dt;
end
diff_nsfmn1 = abs(hilbert(diff_nsfm1)); % Hilbert 变换,求绝对值得到瞬时幅度(包络检波)
zero = (max(diff_nsfmn) - min(diff_nsfmn))/2;
diff_nsfmn1 = diff_nsfmn1 - zero;
subplot(3,1,2);
plot(1:length(diff_nsfm1),diff_nsfm1);           % 绘制含大信噪比高斯白噪声已调信号
                                                 % 的时域图
xlabel('时间 t');
title('含大信噪比高斯白噪声已调信号的时域图');
subplot(3,1,3);                                  % 绘制含大信噪比高斯白噪声解调信号
                                                 % 的时域图
plot((1:length(diff_nsfmn1))./1000,diff_nsfmn1./400,'r');
xlabel('时间 t');
title('含大信噪比高斯白噪声解调信号的时域图');
```

Copy the simulation process of FM without high SNR in the following blank(see Figure A. 17).

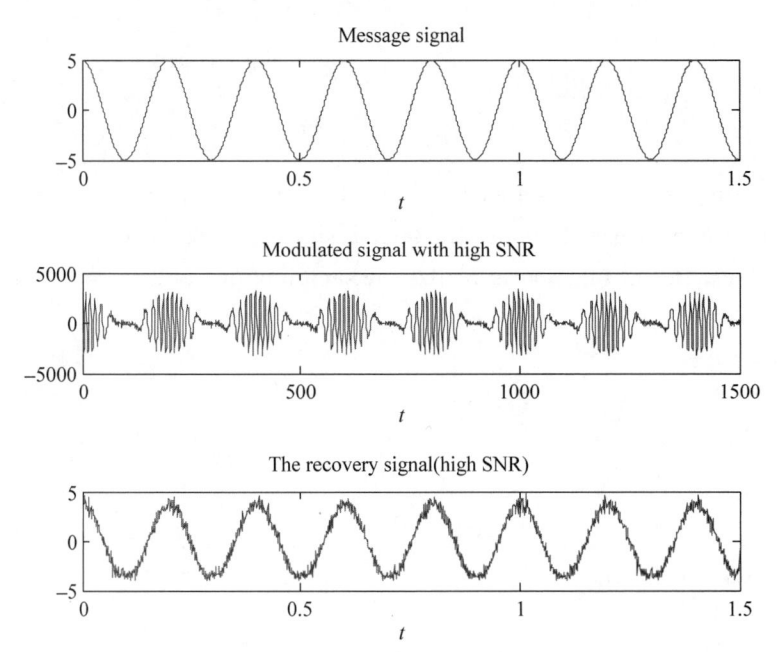

Figure A. 17　The recovery signal with high SNR

4. Discussion the results

Experiment 2　Pulse modulation（corresponding to Chapter 5）

1. Introduction

The principles of the pulse modulation are studied in this experiment，including the sampling process，quantization and the PCM communication systems.

At the end of this experiment，you should have learned：

- the principle of sampling theorem.
- the simulink simulation for sample&hold.
- the PCM process.
- how to carry out simulations using MATLAB software.

2. Pulse modulation

1）Principle of sampling theory

Sampling：an analog signal is converted into a corresponding sequence of samples，as shown in Figure A. 18.

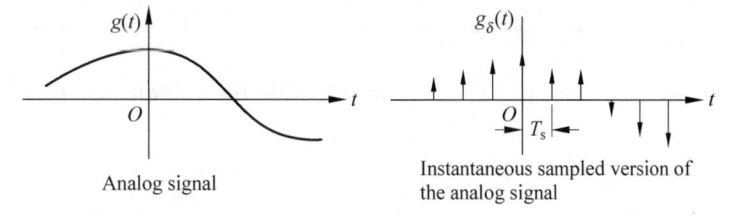

Analog signal

Instantaneous sampled version of the analog signal

Figure A. 18　The sampling process

There are two different methods to get the simulations of the sampling and its reconstruction in MATLAB. Suppose the analog signal is sine wave.

Method 1：use the. m function to realize the sampling process

```
clear;
f = 1000; % 信号频率
fs = 10000; % 采样频率
dt = 0.00001;
t1 = 1/f;ts = 1/fs;
n = 1:400;
m = 1:40;
t = ts * m;
x = sin(2 * pi * f * n * dt);
subplot(311);plot(n/100,x);
title('信号');
xs = sin(2 * pi * f * t);
subplot(312);stem(m/10,xs);
title('采样信号');
%% 设计一个低通的 FIR 滤波器 %%
wp = 2 * 1500/8000;
w = blackman(1 + 1);
```

```
h = fir1(1,wp,'low',w);
xr = filter(h,1,xs);
subplot(313);plot(m/10,xr);
title('重建信号');
```

You can get the simulation of the sampling process. Copy the result in the following blank
(**see Figure A. 19**).

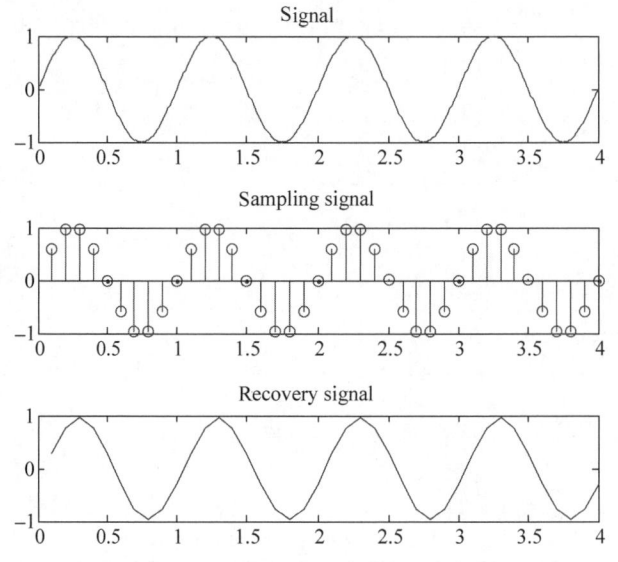

Figure A. 19　The sampling signal and its recovery

Method 2：the sample and hold circuit（Use the simulink to simulate this process）

Create a new model file and then find the following modules，as shown in Figure A. 20.
（You can find these modules from view＞library browser.）

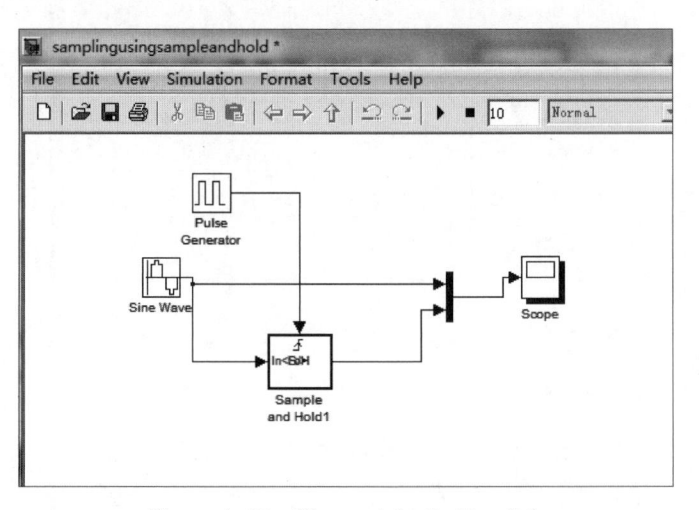

Figure A. 20　The modules in Simulink

The parameter setting of each module is shown in Figure A. 21.

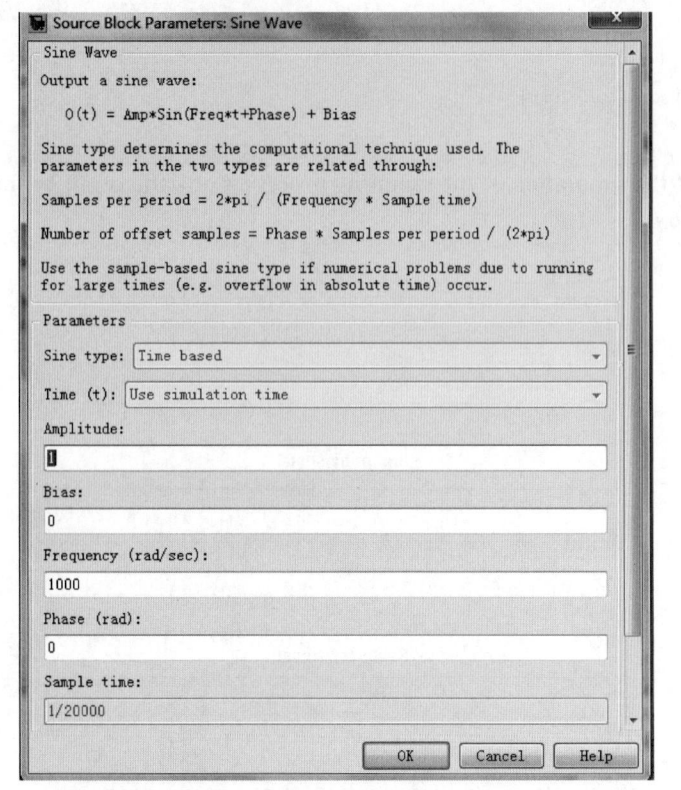

(a) The parameters setting of sine wave

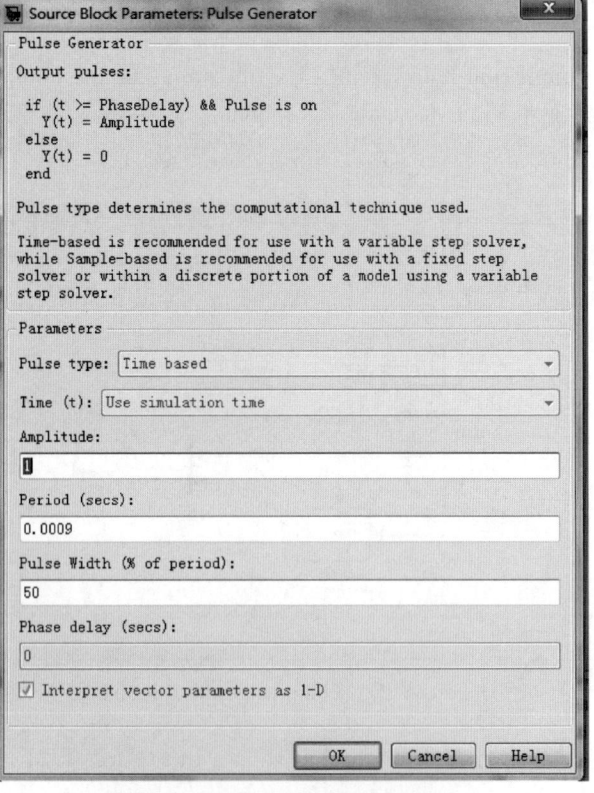

(b) The parameter setting of pulse generator

Figure A.21　The parameters settings

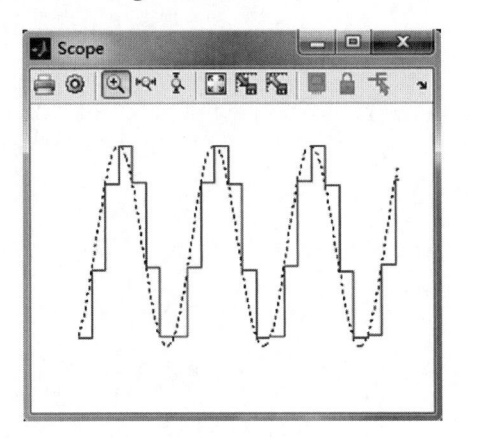

(c) The parameter setting of Mux

Figure A. 21　(continued)

The simulation result is shown in Figure A. 22.

Figure A. 22　The simulation result in simulink

2) The process of PCM system

There are three steps in the process of digitizing an analog signal: sampling, quantization and coding. As can be seen in Figure A. 23, the output signal is PCM.

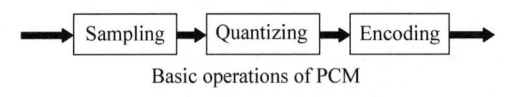

Basic operations of PCM

Figure A. 23　The process of PCM

Step 1: here we also use the sine signal as the original signal

```
clear all;
close all;
% 建立原信号
dt = 0.002;                              % 取时间间隔为 0.01
t = 0:dt:10;                             % 以 dt 为间隔从 0 到 10 画图
fc = 1                                   % xt 里最大频率
xt = sin(2 * pi * fc * t) + cos(2 * pi * fc * t);   % xt 方程
```

```matlab
% 采样: 时间连续信号变为时间离散模拟信号
fs = 10;                                    % 抽样 fs >= 2fc, 每秒内的抽样点数目将大于或等于 2fc 个
sdt = 1/fs;                                 % 频域采样间隔 0.1
t1 = 0 : sdt : 10;                          % 以 sdt 为间隔从 0 到 10 画图
st = sin(2 * pi * fc * t1) + cos(2 * pi * fc * t1);   % coswt = cos2pift, 2pif = w
figure(1);
subplot(311);
plot(t, xt); title('原始信号');            % 条状图, 连续图
grid on                                     % 画背景
subplot(312);
stem(t1, st, '.');                          % 杆状图, 离散图
title('抽样信号');
grid on                                     % 画背景


% 量化编码一步完成
n = length(st);                             % 取 st 的长度为 n, 本题函数 n 为 101
    M = max(st);
    A = (st/M) * 2048;
% a1(极性码) a2a3a4(段落码)a5a6a7a8(段内电平码)
    code = zeros(n, 8);                     % 产生 i * 8 的零矩阵
% 极性码 a1
for i = 1 : n                               % if 循环语句

    if A(i) >= 0
        code(i, 1) = 1;                     % 代表正值
    else
        code(i, 1) = 0;                     % 代表负值
    end


% 段内码 a2a3a4
if abs(A(i)) >= 0 && abs(A(i)) < 16
        code(i, 2) = 0; code(i, 3) = 0; code(i, 4) = 0; step = 1; start = 0;
elseif 16 <= abs(A(i)) && abs(A(i)) < 32
        code(i, 2) = 0; code(i, 3) = 0; code(i, 4) = 1; step = 1; start = 16;
elseif 32 <= abs(A(i)) && abs(A(i)) < 64
        code(i, 2) = 0; code(i, 3) = 1; code(i, 4) = 0; step = 2; start = 32;
elseif 64 <= abs(A(i)) && abs(A(i)) < 128
        code(i, 2) = 0; code(i, 3) = 1; code(i, 4) = 1; step = 4; start = 64;
elseif 128 <= abs(A(i)) && abs(A(i)) < 256
        code(i, 2) = 1; code(i, 3) = 0; code(i, 4) = 0; step = 8; start = 128;
elseif 256 <= abs(A(i)) && abs(A(i)) < 512
        code(i, 2) = 1; code(i, 3) = 0; code(i, 4) = 1; step = 16; start = 256;
elseif 512 <= abs(A(i)) && abs(A(i)) < 1024
        code(i, 2) = 1; code(i, 3) = 1; code(i, 4) = 0; step = 32; start = 512;
elseif 1024 <= abs(A(i)) && abs(A(i)) < 2048
        code(i, 2) = 1; code(i, 3) = 1; code(i, 4) = 1; step = 64; start = 1024;
end
B = floor((abs(A(i)) - start)/step);        % 段内码编码 floor 取整 4 舍 5 入

    t = dec2bin(B, 4) - 48;                 % dec2bin 定义将 B 变为 4 位二进制码, -48 改变格式
```

```
        code(i,5:8) = t(1:4);              % 输出段内码
end
code = reshape(code',1,8 * n);             % reshape 代表重新塑形
code
subplot(313);
stem(code,'.');axis([1 100 0 1]);
title('编码信号');
grid on
```

Save these code as. m file，and run it. Copy the result in the following blank(see Figure A. 24).

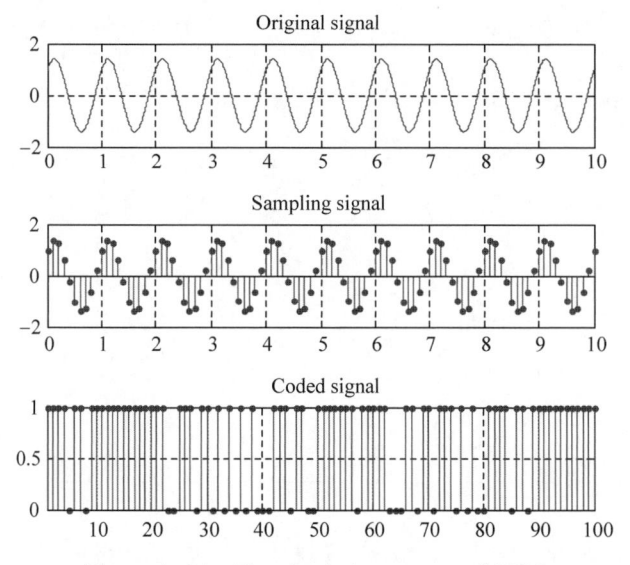

Figure A. 24　The simulation process of PCM

Step 2：we change the sine wave to the speech signal. The speech file is test1. wav. Open it in MATLAB using wavread function(see Figure A. 25)

```
[[x,fs,bits] = wavread('test1.wav');
subplot(211);plot(x);title('抽样信号');
n = length(x);
            M = max(x);
    A = (x/M) * 2048;
code = zeros(i,8);
for i = 1:n
    if A(i)> = 0
        code(i,1) = 1;
    else
        code(i,1) = 0;
    end
if abs(A(i))> = 0&&abs(A(i))< 16
        code(i,2) = 0;code(i,3) = 0;code(i,4) = 0;step = 1;start = 0;
elseif 16 < = abs(A(i))&&abs(A(i))< 32
        code(i,2) = 0;code(i,3) = 0;code(i,4) = 1;step = 1;start = 16;
elseif 32 < = abs(A(i))&&abs(A(i))< 64
```

```
        code(i,2) = 0;code(i,3) = 1;code(i,4) = 0;step = 2;start = 32;
elseif 64 < = abs(A(i))&&abs(A(i))< 128
        code(i,2) = 0;code(i,3) = 1;code(i,4) = 1;step = 4;start = 64;
elseif 128 < = abs(A(i))&&abs(A(i))< 256
        code(i,2) = 1;code(i,3) = 0;code(i,4) = 0;step = 8;start = 128;
elseif 256 < = abs(A(i))&&abs(A(i))< 512
        code(i,2) = 1;code(i,3) = 0;code(i,4) = 1;step = 16;start = 256;
elseif 512 < = abs(A(i))&&abs(A(i))< 1024
        code(i,2) = 1;code(i,3) = 1;code(i,4) = 0;step = 32;start = 512;
elseif 1024 < = abs(A(i))&&abs(A(i))< 2048
        code(i,2) = 1;code(i,3) = 1;code(i,4) = 1;step = 64;start = 1024;
end
B = floor((abs(A(i)) - start)/step);
    t = dec2bin(B,4) - 48;
    code(i,5:8) = t(1:4);
end
code = reshape(code',1,8 * n);
code
subplot(212);
stem(code,'.');axis([300000 300500 0 1]);
title('编码信号');
grid on
```

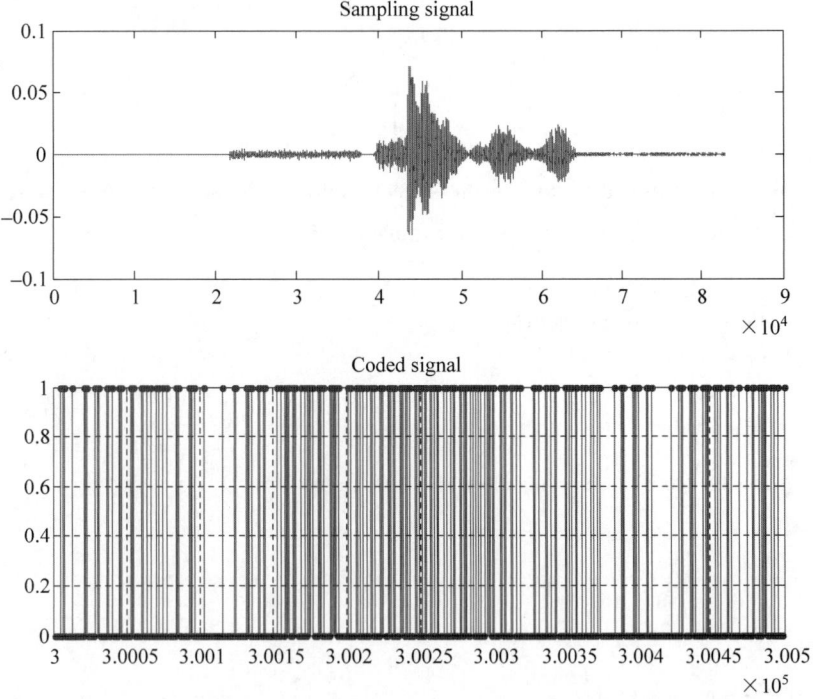

Figure A. 25　The PCM signal of voice signal

Step 3：the encode and decode process of PCM

The principle is as following flow chart Figure A.26.

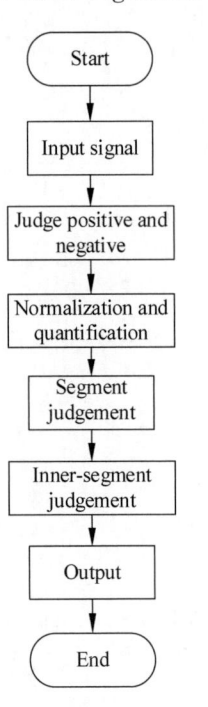

Figure A.26　The flow chart of PCM

The encode function is saved as pcm.m

```
% PCM 编码程序
function code = pcm(S)                              % S 为输入信号
z = sign(S);                                        % 判断 S 的正负
MaxS = max(abs(S));                                 % 求 S 的最大值
S = abs(S/MaxS);                                    % 归一化
Q = 2048 * S;                                       % 量化
code = zeros(length(S),8);                          % 代码存储矩阵(全零)
% 段落码判断程序
for i = 1:length(S)
    if (Q(i)> = 128)&(Q(i)< = 2048)
        code(i,2) = 1;                              % 在第五段与第八段之间,段位码第一位都为"1"
    end
    if (Q(i)> 32)&(Q(i)< 128)||(Q(i)> = 512)&(Q(i)< = 2048)
        code(i,3) = 1;                              % 在第三、四、七、八段内,段位码第二位为"1"
    end
    if (Q(i)> = 16)&(Q(i)< 32)||(Q(i)> = 64)&(Q(i)< 128)||(Q(i)> = 256)&(Q(i)< 512)||(Q(i)> =
1024)&(Q(i)< = 2048)
        code(i,4) = 1;                              % 在第二、四、六、八段内,段位码第三位为"1"
    end
end
% 段内码判断程序
N = zeros(length(S));
for i = 1:length(S)
N(i) = bin2dec(num2str(code(i,2:4))) + 1;           % 找到 code 位于第几段
```

```
end
a = [0,16,32,64,128,256,512,1024];              % 量化间隔
b = [1,1,2,4,8,16,32,64];                        % 除以 16,得到每段的最小量化间隔
for i = 1:length(S)
    q = ceil((Q(i) - a(N(i)))/b(N(i)));          % 求出在段内的位置
    if q == 0
        code(i,(5:8)) = [0,0,0,0];               % 如果输入为零,则输出"0"
    else k = num2str(dec2bin(q-1,4));            % 编码段内码为二进制
    code(i,5) = str2num(k(1));
    code(i,6) = str2num(k(2));
    code(i,7) = str2num(k(3));
    code(i,8) = str2num(k(4));
    end
    if z(i)> 0
        code(i,1) = 1;
    elseif z(i)< 0
        code(i,1) = 0;
    end                                          % 符号位的判断
end
```

You can test it use a data k＝1270 or 635. And then k is as the input data of this pcm function in the command window.

```
Test this process
>> k = [2048 1270 635]
    a = pcm(k)
Then try to get the value of a.
```

Step 4：Use the simulink to simulate sampling process

Create a new model file and then find the following modules as shown in Figure A. 27.

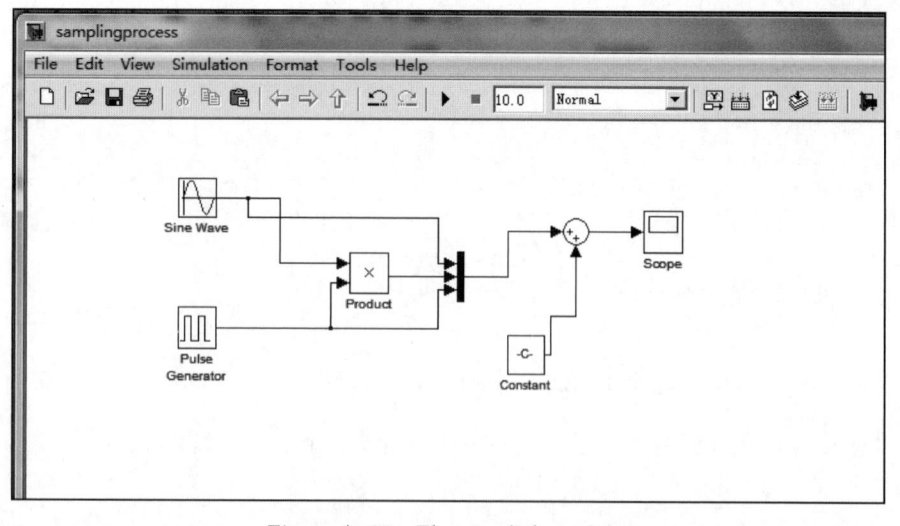

Figure A. 27　The simulink modules

From View-library browser to find the sine wave and other modules，and the parameter settings are shown in Figure A. 28.

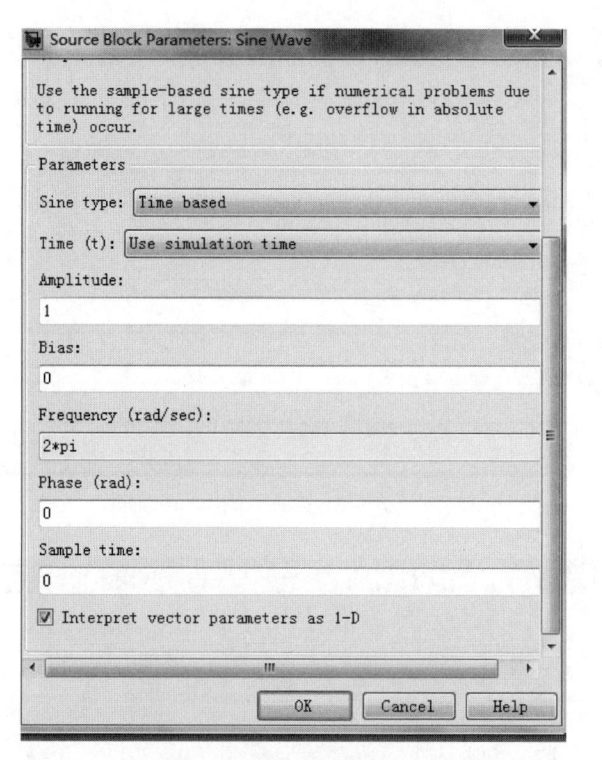

(a) The parameters setting of sine wave

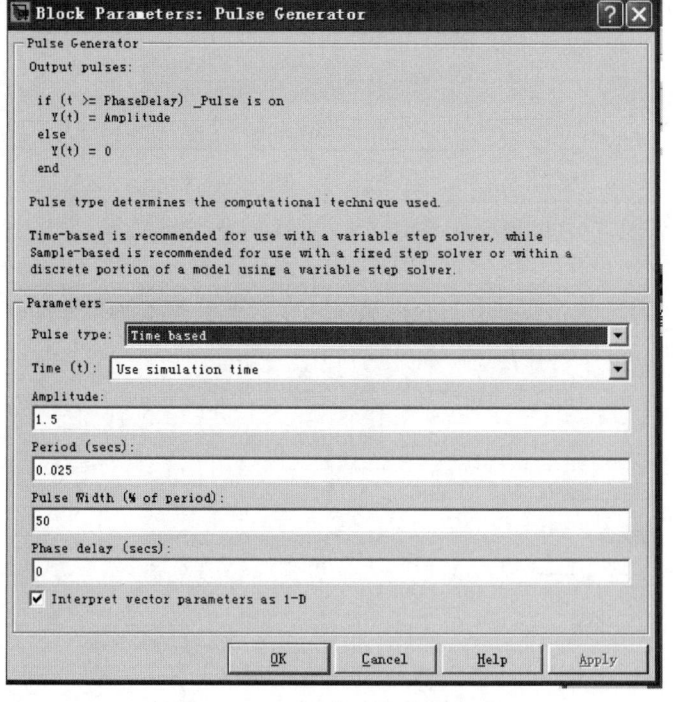

(b) The parameters setting of pulse generator

Figure A. 28　The parameters setting in each module

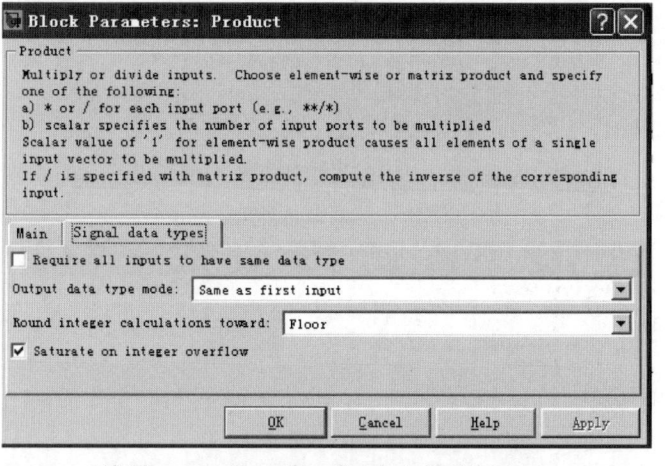

(c) The parameters setting of product-main

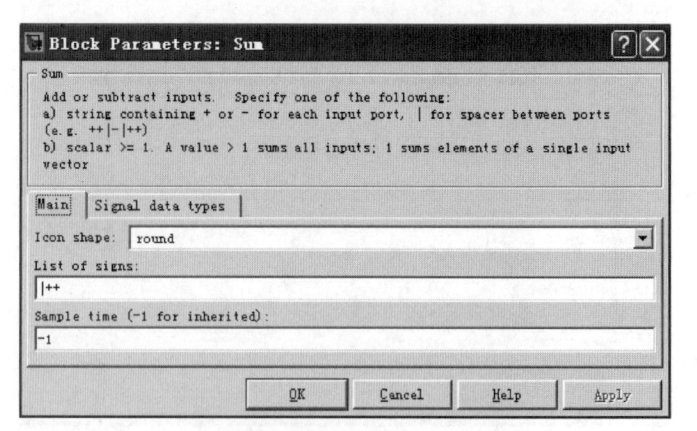

(d) The parameters setting of product-signal data types

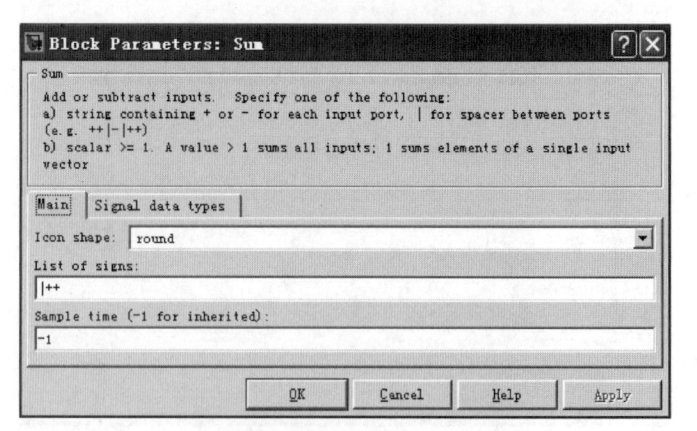

(e) The parameters setting of sum-main

Figure A.28　（continued）

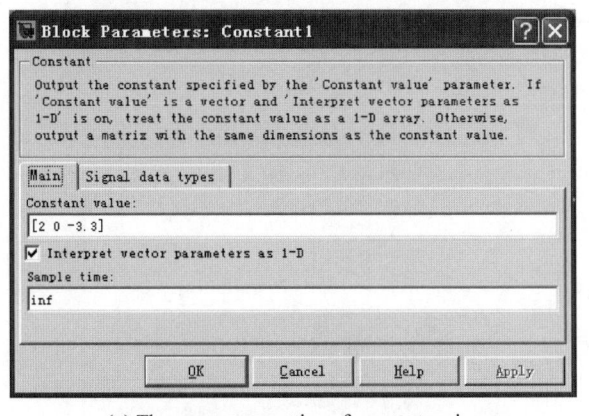

(f) The parameters setting of sum- signal data types

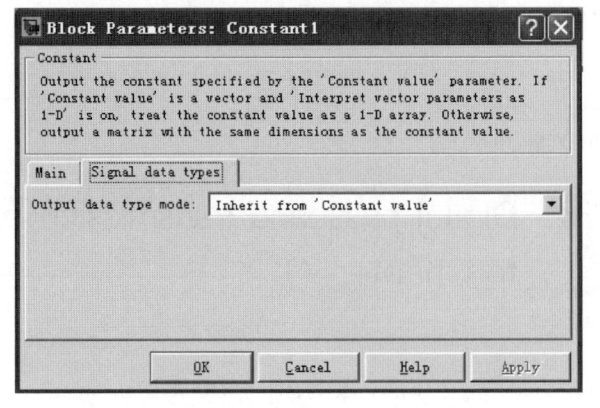

(g) The parameters setting of constant-main

(h) The parameters setting of constant- signal data types

Figure A. 28　(continued)

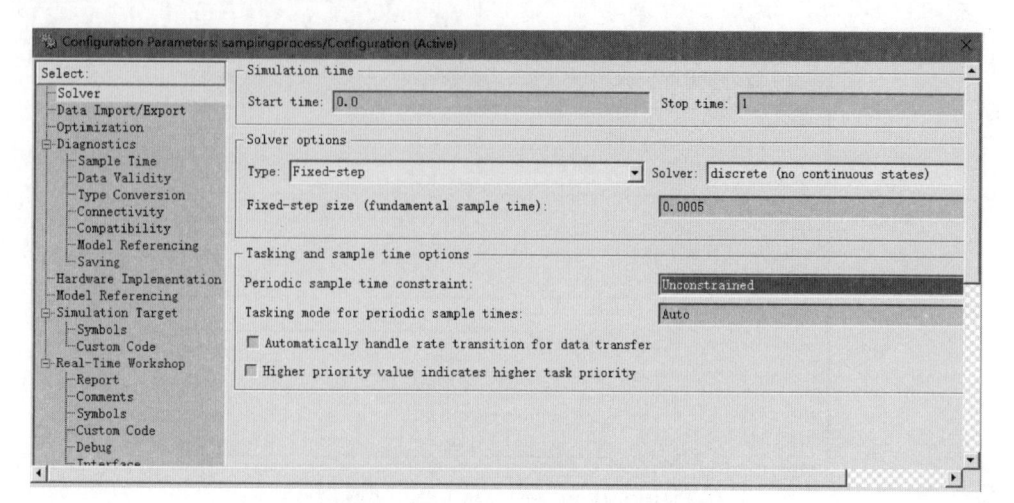

(i) The block and configuration parameters setting

Figure A. 28　（continued）

The simulation result is shown in the following Figure A. 29.

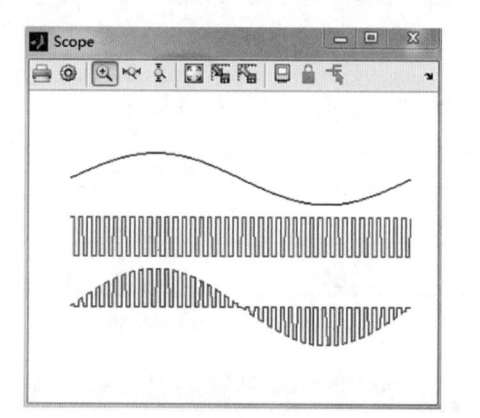

Figure A. 29　The simulation result of sampling process

3. Discussion the results

Experiment 3　Digital passband transmission（corresponding to Chapter 7）

1. Introduction

The generation and detection of ASK/FSK/PSK will be studied in this part. At the end of this experiment，you should have learned：

- the principle of generation and detection of digital passband transmission.
- the simulation of AMI& HDB$_3$ encoder.
- how to carry out simulations using MATLAB software.

2. Passband digital modulation and demodulation

Functional model of passband data transmission system is shown in Figure A. 30.

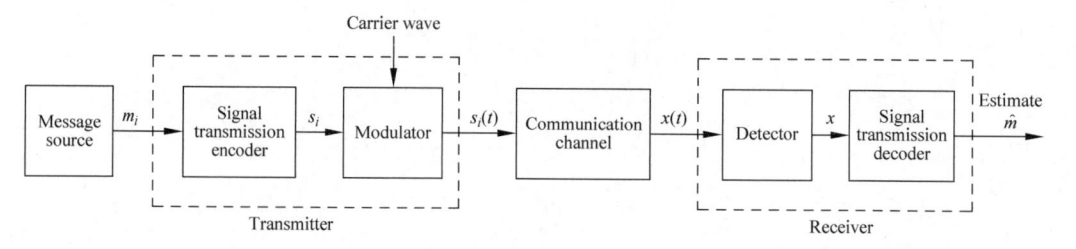

Figure A. 30 Model of passband digital transmission system

1) Principle of ASK (amplitude-shift keying)

The block diagrams of the generation and coherent detection of ASK are shown in Figure A. 31 and Figure A. 32.

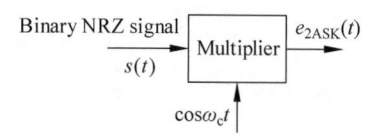

Figure A. 31   The generation of ASK

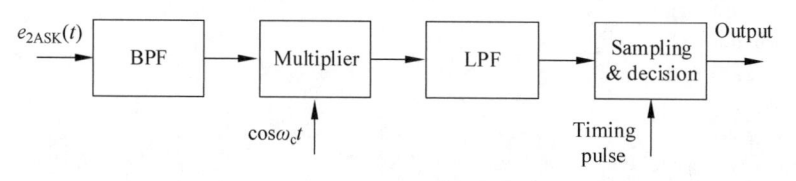

Figure A. 32   The coherent detection of ASK

The MATLAB simulations code of the generation and detection of ASK are in the following. Suppose the carrier wave is sine wave. Use the main. m function to realize the whole process.

```
clear all
close all
i = 10;
j = 5000;
t = linspace(0,5,j); % 取 0,0.001,0.002, … 共 5001 个点
fc = 10;
fm = i/5;
B = 2 * fc;
%%%%%%%%%%%%%%%%%%%%% 产生基带信号
a = round(rand(1,i));                    % 随机序列
figure(2)
plot(rand(1,i))
st = t;
for n = 1:10
    if a(n)< 1;
```

```
        for m = j/i * (n - 1) + 1:j/i * n          % 一个信号取 500 个点
            st(m) = 0;
        end
    else
        for m = j/i * (n - 1) + 1:j/i * n
            st(m) = 1;
        end
    end
end
figure(1);
subplot(421);
plot(t, st);
axis([0, 5, -1, 2]);
title('基带信号');
%%%%%%%%%%%%%%%%%%%%%%%% 载波
s1 = cos(2 * pi * fc * t);
subplot(422);
plot(s1);
title('载波信号');
%%%%%%%%%%%%%%%%%%%%%%%%% 调制
e_2ask = st. * s1;
subplot(423);
plot(t, e_2ask);
title('已调信号');
noise = rand(1, j);
e_2ask = e_2ask + noise;                   % 加入噪声
subplot(424);
plot(t, e_2ask);
title('加入噪声的信号');
%%%%%%%%%%%%%%%%%%%%%%% 相干解调
at = e_2ask. * cos(2 * pi * fc * t);
at = at - mean(at);
subplot(425);
plot(t, at);
title('相乘后信号');
[f, af] = T2F(t, at);    % 通过低通滤波器
[t, at] = lpf(f, af, 2 * fm);
subplot(426);
plot(t, at);
title('解调后波形');
%%%%%%%%%%%%%%%%%%%%%%%%%%%%%% 抽样判决
for m = 0:i - 1;
if at(1, m * 500 + 250) + 0.5 < 0.5;
    for j = m * 500 + 1:(m + 1) * 500;
        at(1, j) = 0;
    end
else
    for j = m * 500 + 1:(m + 1) * 500;
        at(1, j) = 1;
    end
end
```

```
end
subplot(427);
plot(t,at);
axis([0,5,-1,2]);
title('抽样判决后波形')
```

The sub-function T2F is as following（create a new. m function）

```
function [f,sf] = T2F(t,st)
dt = t(2) - t(1);
T = t(end);
df = 1/T;
N = length(st);
f = -N/2 * df:df:N/2 * df - df;
sf = fft(st);
sf = T/N * fftshift(sf);
```

The sub-function F2T is as following：

```
function [t,st] = F2T(f,sf)
% This function calculates the time signal using ifft function for the input
% signal's spectrum
df = f(2) - f(1);
Fmx = ( f(end) - f(1) + df);
dt = 1/Fmx;
N = length(sf);
T = dt * N;
% t =- T/2:dt:T/2 - dt;
t = 0:dt:T - dt;
sff = fftshift(sf);
st = Fmx * ifft(sff);
```

The low-pass filter function is as following：

```
function [t st] = lpf(f,sf,B)
% This function filters an input data using a lowpass filter
% Inputs: f: frequency samples
% sf: input data spectrum samples
% B: lowpass's bandwidth with a rectangle lowpass
% Outputs: t: time samples
% st: output data's time samples
df = f(2) - f(1);
T = 1/df;
hf = zeros(1,length(f));
bf = [-floor( B/df ): floor( B/df )] + floor( length(f)/2 );
hf(bf) = 1;
yf = hf. * sf;
[t,st] = F2T(f,yf);
st = real(st);
```

You can get the simulations of the ASK process. Copy the results in the following blank (see Figure A. 33).

2) Principle of FSK (frequency-shift keying)

The symbols 1 and 0 of 2FSK are transmitted respectively with two different carrier

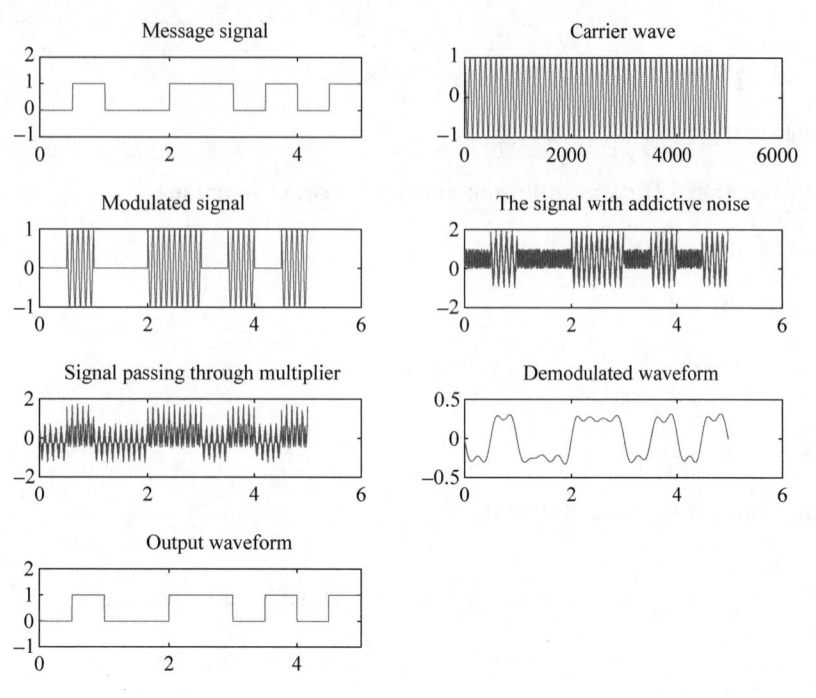

Figure A. 33　The simulation of ASK process

frequencies. 2FSK is equivalence of two 2ASK's superposition (叠加). The generation and detection are shown in Figure A. 34.

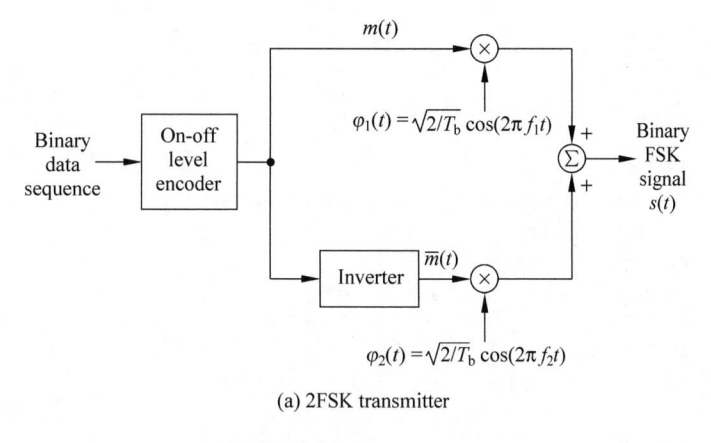

(a) 2FSK transmitter

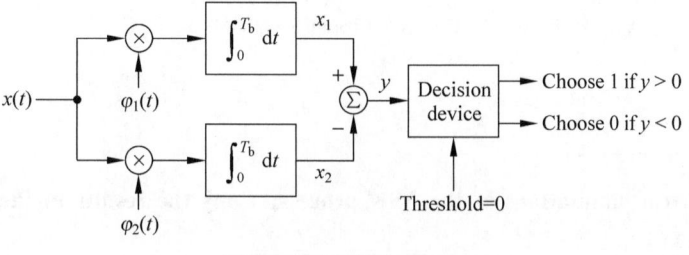

(b) Coherent 2FSK receiver

Figure A. 34　The generation and coherent detection of 2FSK

The MATLAB simulations code of the generation and detection of FSK are in the following. Use the. m function to realize the whole process.

```
clear all
close all
i = 10; % 基带信号码元数
j = 5000;
a = round(rand(1,i)); % 产生随机序列
t = linspace(0,5,j);
f1 = 10; % 载波 1 频率
f2 = 5; % 载波 2 频率
fm = i/5; % 基带信号频率
B1 = 2 * f1; % 载波 1 带宽
B2 = 2 * f2; % 载波 2 带宽
%%%%%%%%%%%%%%%%%% 产生基带信号
st1 = t;
for n = 1:10
    if a(n) < 1;
        for m = j/i * (n - 1) + 1:j/i * n
            st1(m) = 0;
        end
    else
        for m = j/i * (n - 1) + 1:j/i * n
            st1(m) = 1;
        end
    end
end
st2 = t;
%%%%%%%%%%%%%%%%%% 基带信号求反
for n = 1:j;
    if st1(n) >= 1;
        st2(n) = 0;
    else
        st2(n) = 1;
    end
end;
figure(1);
subplot(411);
plot(t,st1);
title('基带信号');
axis([0,5, - 1,2]);
subplot(412);
plot(t,st2);
title('基带信号反码');
axis([0,5, - 1,2]);
%%%%%%%%%%%%%%%%%%%%%%% 载波信号
s1 = cos(2 * pi * f1 * t)
s2 = cos(2 * pi * f2 * t)
```

```
subplot(413),plot(s1);
title('载波信号 1');
subplot(414),plot(s2);
title('载波信号 2');
%%%%%%%%%%%%%%%%%%%%%%%% 调制
F1 = st1. * s1; % 加入载波 1
F2 = st2. * s2; % 加入载波 2
figure(2);
subplot(411);
plot(t,F1);
title('s1 * st1');
subplot(412);
plot(t,F2);
title('s2 * st2');
e_fsk = F1 + F2;
subplot(413);
plot(t,e_fsk);
title('2FSK 信号')
noise = rand(1,j);
fsk = e_fsk + noise;
subplot(414);
plot(t,fsk);
title('加噪声信号')
%%%%%%%%%%%%%%%%%%%%%%%% 相干解调
st1 = fsk. * s1; % 与载波 1 相乘
[f,sf1] = T2F(t,st1); % 通过低通滤波器
[t,st1] = lpf(f,sf1,2 * fm);
figure(3);
subplot(311);
plot(t,st1);
title('与载波 1 相乘后波形');
st2 = fsk. * s2; % 与载波 2 相乘
[f,sf2] = T2F(t,st2); % 通过低通滤波器
[t,st2] = lpf(f,sf2,2 * fm);
subplot(312);
plot(t,st2);
title('与载波 2 相乘后波形');
for m = 0:i-1;
%%%%%%%%%%%%%%%%%%%%%%%% 抽样判决
if st1(1,m * 500 + 250)< 0.25;
for j = m * 500 + 1:(m + 1) * 500;
at(1,j) = 0;
end
else
for j = m * 500 + 1:(m + 1) * 500;
at(1,j) = 1;
end
end
end;
subplot(313);
plot(t,at);
```

```
axis([0,5,-1,2]);
title('抽样判决后波形')
```

You can get the simulations of the FSK process. Copy the results in the following blank (see Figure A. 35 ~ Figure A. 37).

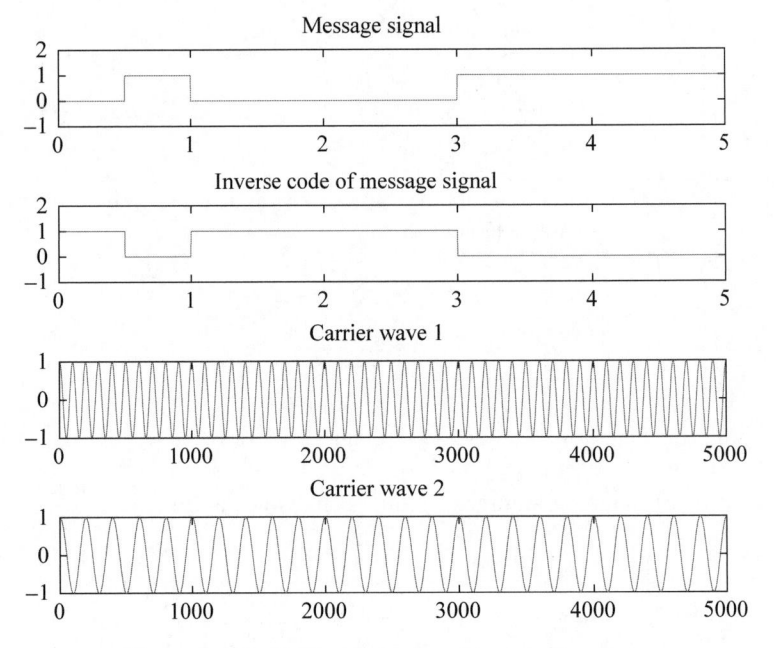

Figure A. 35   Baseband signal and two carrier waves

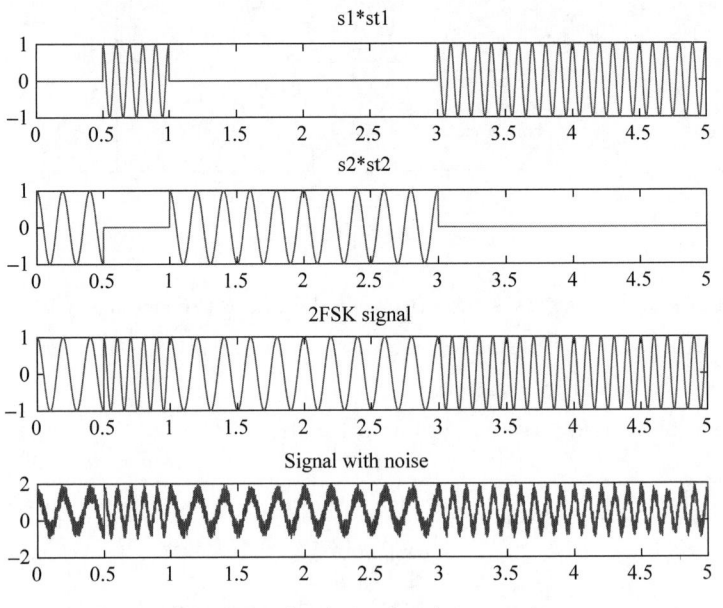

Figure A. 36   Baseband signal and two carrier waves

3）Principle of QPSK

The **QPSK** Modulator uses a bit-splitter，two multipliers with local oscillator，a 2-bit

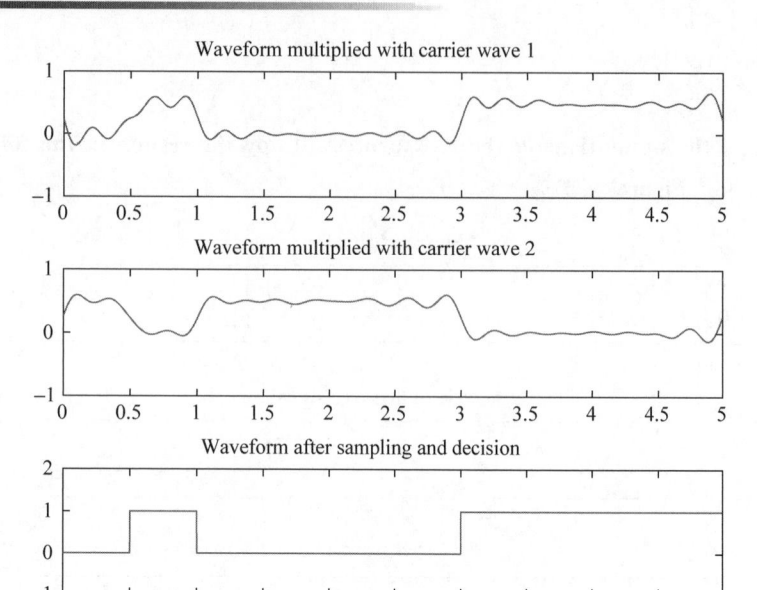

Figure A. 37   Baseband signal and two carrier waves

serial to parallel converter，and a summer circuit. Following Figure A. 38 is the block diagram.

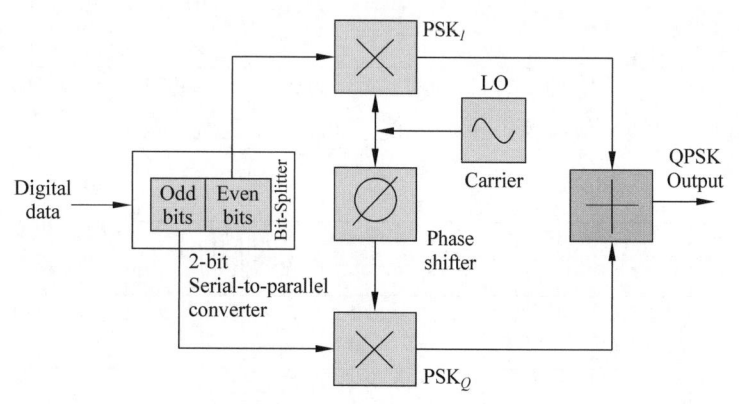

Figure A. 38   QPSK modulator

The QPSK demodulator uses two product demodulator circuits with local oscillator，two band pass filters，two integrator circuits，and a 2-bit parallel to serial converter. Following Figure A. 39 is the diagram.

（1）Create a new. model file as the main function.

```
%% 初始化参数
%%%%%%%%%%%%%%%%%%%%%%
T = 1;                      % 基带信号宽度,也就是频率
fc = 10/T;                  % 载波频率
ml = 2;                     % 调制信号类型的一个标志位
nb = 100;                   % 传输的比特数
delta_T = T/200;            % 采样间隔
fs = 1/delta_T;             % 采样频率
SNR = 0;                    % 信噪比
```

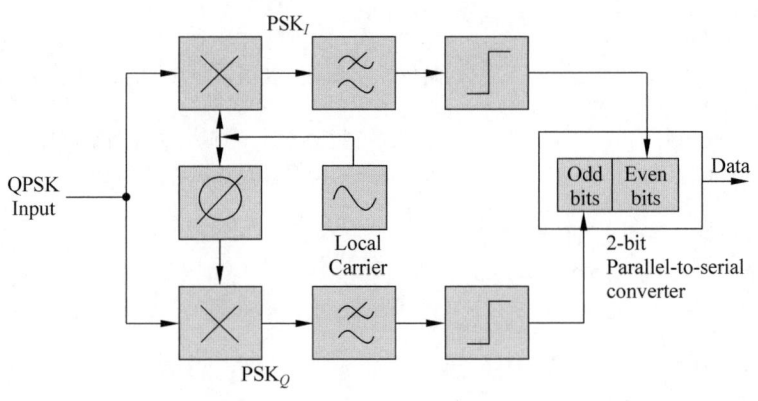

Figure A. 39   The demodulator of QPSK

```
t = 0:delta_T:nb * T - delta_T;        % 限定 t 的取值范围
N = length(t);                         % 采样数
%%%%%%%%%%%%%%%%%%%%%%%%
  %% 调制部分
%%%%%%%%%%%%%%%%%%%%%%
% 基带信号的产生
data = randn(1,nb)> 0.5;               % 调用一个随机函数(0 or 1),输出到一个 1 * 100 的矩阵
datanrz = data. * 2 - 1;               % 变成极性码
data1 = zeros(1,nb/delta_T);           % 创建一个 1 * nb/delta_T 的零矩阵
for q = 1:nb
    data1((q - 1)/delta_T + 1:q/delta_T) = datanrz(q); % 将极性码变成对应的波形信号
end
% 将基带信号变换成对应波形信号
data0 = zeros(1,nb/delta_T);           % 创建一个 1 * nb/delta_T 的零矩阵
for q = 1:nb
    data0((q - 1)/delta_T + 1:q/delta_T) = data(q);         % 将极性码变成对应的波形信号
end
% 发射的信号
data2 = abs(fft(data1));
% 串并转换,将奇偶位数据分开
idata = datanrz(1:ml:(nb - 1));        % 将奇偶位分开,因此间隔 m1 为 2
qdata = datanrz(2:ml:nb);
% QPSK 信号的调制
ich = zeros(1,nb/delta_T/2);           % 创建一个 1 * nb/delta_T/2 的零矩阵,以便后面存放奇偶位数据
for i = 1:nb/2
    ich((i - 1)/delta_T + 1:i/delta_T) = idata(i);
end
for ii = 1:N/2
    a(ii) = sqrt(2/T) * cos(2 * pi * fc * t(ii));
end
idata1 = ich. * a;                     % 奇数位数据与余弦函数相乘,得到一路的调制信号
qch = zeros(1,nb/2/delta_T);
for j1 = 1:nb/2
    qch((j1 - 1)/delta_T + 1:j1/delta_T) = qdata(j1);
end
for jj = 1:N/2
```

```
        b(jj) = sqrt(2/T) * sin(2 * pi * fc * t(jj));
    end
    qdata1 = qch. * b;                        % 偶数位数据与余弦函数相乘,得到另一路的调制信号

    s = idata1 + qdata1;                      % 将奇偶位数据合并,s 即为 QPSK 调制信号
    ss = abs(fft(s));                         % 快速傅里叶变换得到频谱
    %%%%%%%%%%%%%%%%%%%
       %% 瑞利衰落信道和高斯信道
    %%%%%%%%%%%%%%%%%%%
    % 瑞利衰落信道
    ray_ich = raylrnd(0.8,1,nb/2/delta_T);
    ray_qch = raylrnd(0.8,1,nb/2/delta_T);
    Ray_idata = idata1. * ray_ich;
    Ray_qdata = qdata1. * ray_qch;
    Ray_s = Ray_idata + Ray_qdata;
    %%%%%%%%%%%%%%%%%
    % 高斯信道
    s1 = awgn(s,SNR);                         % 通过高斯信道之后的信号
    s11 = abs(fft(s1));                       % 快速傅里叶变换得到频谱
    s111 = s1 - s;                            % 高斯噪声曲线
    %%%%%%%%%%%%%%%%%%
    Awgn_s = awgn(Ray_s,SNR);                 % 通过高斯信道再通过瑞利衰落信道
    %%%%%%%%%%%%%%%%%%%%
    %% QPSK 解调部分
    %%%%%%%%%%%%%%%%%%%
    % 解调部分(高斯信道)
    idata2 = s1. * a;                         % 这里面其实隐藏了一个串并转换的过程
    qdata2 = s1. * b;                         % 对应的信号与正余弦信号相乘
    idata3 = zeros(1,nb/2);                   % 建立 1 * nb 数组,以存放解调之后的信号
    qdata3 = zeros(1,nb/2);
    % 抽样判决的过程,与 0 作比较,data >= 0,则置 1,否则置 0
    for n = 1:nb/2
        A1(n) = sum(idata2((n-1)/delta_T + 1:n/delta_T));
        if sum(idata2((n-1)/delta_T + 1:n/delta_T)) >= 0
            idata3(n) = 1;
        else idata3(n) = 0;
        end
        A2(n) = sum(qdata2((n-1)/delta_T + 1:n/delta_T));
        if sum(qdata2((n-1)/delta_T + 1:n/delta_T)) >= 0
            qdata3(n) = 1;
        else qdata3(n) = 0;
        end
    end
    % 为了显示星座图,将信号进行处理
    idata4 = zeros(1,nb/2);
    qdata4 = zeros(1,nb/2);
    for n = 1:nb/2
        Awgn_ichsum(n) = sum(idata2((n-1)/delta_T + 1:n/delta_T)) * delta_T;
        if Awgn_ichsum(n) >= 0
            idata4(n) = 1;
        else idata4(n) = 0;
```

```
    end
    Awgn_qchsum(n) = sum(qdata2((n - 1)/delta_T + 1:n/delta_T)) * delta_T;
    if Awgn_qchsum(n)> = 0
        qdata4(n) = 1;
    else qdata4(n) = 0;
    end
end
% 将判决之后的数据存放进数组
demodata = zeros(1,nb);
demodata(1:ml:(nb - 1)) = idata3;     % 存放奇数位
demodata(2:ml:nb) = qdata3;           % 存放偶数位
% 为了显示,将它变成波形信号(即传输一个 1 代表单位宽度的高电平)
demodata1 = zeros(1,nb/delta_T);       % 创建一个 1 * nb/delta_T 的零矩阵
for q = 1:nb
    demodata1((q - 1)/delta_T + 1:q/delta_T) = demodata(q);  % 将极性码变成对应的波形信号
end
% 累计误码数
% abs(demodata - data)求接收端和发射端
% 数据差的绝对值,累计之后就是误码个数
Awgn_num_BER = sum(abs(demodata - data))
%%%%%%%%%%%%%%%%%%%%
% 解调部分(瑞利 + 高斯)
Ray_idata2 = Ray_s. * a;              % 这里面其实隐藏了一个串并转换的过程
Ray_qdata2 = Ray_s. * b;              % 对应的信号与正余弦信号相乘
% Ray_idata3 = zeros(1,nb/2);          % 建立 1 * nb 数组,以存放解调之后的信号
% Ray_qdata3 = zeros(1,nb/2);
% 抽样判决的过程,与 0 作比较,data > = 0,则置 1,否则置 0
% for n = 1:nb/2
%     if Ray_sum(Ray_idata2((n - 1)/delta_T + 1:n/delta_T))> = 0
%         Ray_idata3(n) = 1;
%     else Ray_idata3(n) = 0;
%     end
%     if Ray_sum(Ray_qdata2((n - 1)/delta_T + 1:n/delta_T))> = 0
%         Ray_qdata3(n) = 1;
%     else Ray_qdata3(n) = 0;
%     end
% end
% 为了显示星座图,将信号进行处理
Ray_idata4 = zeros(1,nb/2);
Ray_qdata4 = zeros(1,nb/2);
for n = 1:nb/2
    Ray_ichsum(n) = sum(idata2((n - 1)/delta_T + 1:n/delta_T)) * delta_T;
    if Ray_ichsum(n)> = 0
        Ray_idata4(n) = 1;
    else Ray_idata4(n) = 0;
    end
    Ray_qchsum(n) = sum(qdata2((n - 1)/delta_T + 1:n/delta_T)) * delta_T;
    if Ray_qchsum(n)> = 0
        Ray_qdata4(n) = 1;
    else Ray_qdata4(n) = 0;
    end
```

```
end

    % 将判决之后的数据存放进数组
Ray_demodata = zeros(1,nb);
Ray_demodata(1:ml:(nb-1)) = Ray_idata4;      % 存放奇数位
Ray_demodata(2:ml:nb) = Ray_qdata4;          % 存放偶数位
    % 为了显示,将它变成波形信号(即传输一个 1 代表单位宽度的高电平)
Ray_demodata1 = zeros(1,nb/delta_T);         % 创建一个 1 * nb/delta_T 的零矩阵
for q = 1:nb
    Ray_demodata1((q-1)/delta_T+1:q/delta_T) = Ray_demodata(q); % 将极性码变成对应的波形信号
end                                          % 累计误码数
    % abs(demodata-data)求接收端和发射端
    % 数据差的绝对值,累计之后就是误码个数
Ray_num_BER = sum(abs(Ray_demodata - data))


    %%%%%%%%%%%%%%%%%%
    %%                  误码率计算
    %%        调用了 cm_sm32()和 cm_sm33()函数
    %% 声明：函数声明在另外两个 M 文件中
    %% 作用：cm_sm32()用于瑞利信道误码率的计算
    %%        cm_sm33()用于高斯信道误码率的计算
    %% echo on/off 作用在于决定是否显示指令内容
    %%%%%%%%%%%%%%%%%%
SNRindB1 = 0:1:6;
SNRindB2 = 0:0.1:6;
    % 瑞利衰落信道
for i = 1:length(SNRindB1),
    [pb,ps] = cm_sm32(SNRindB1(i));          % 比特误码率
    smld_bit_ray_err_prb(i) = pb;
    smld_symbol_ray_err_prb(i) = ps;
    disp([ps,pb]);
    echo off;
end;
    % 高斯信道
echo on;
for i = 1:length(SNRindB1),
    [pb1,ps1] = cm_sm33(SNRindB1(i));
    smld_bit_awgn_err_prb(i) = pb1;
    smld_symbol_awgn_err_prb(i) = ps1;
    disp([ps1,pb1]);
    echo off;
end;
    % 理论曲线
echo on;
for i = 1:length(SNRindB2),
    SNR = exp(SNRindB2(i) * log(10)/10);             % 信噪比
    theo_err_awgn_prb(i) = 0.5 * erfc(sqrt(SNR));     % 高斯噪声理论误码率
    theo_err_ray_prb(i) = 0.5 * (1 - 1/sqrt(1 + 1/SNR));% 瑞利衰落信道理论误码率
    echo off;
end;
    %%%%%%%%%%%%%%%%%%%
```

```matlab
h = spectrum.welch;                          % 类似于 C 语言的宏定义,方便以下的调用
%%%%%%%%%%%%%%%%%
输出显示部分
%%%%%%%%%%%%%%%
% 第一部分(理想)
figure(1)
subplot(3,2,1);
plot(data0),title('基带信号');
axis([0 20000 - 2 2]);
subplot(3,2,2);
psd(h,data1,'fs',fs),title('基带信号功率谱密度');
subplot(3,2,3);
plot(s),title('调制信号');
axis([200 800 - 3 3]);
subplot(3,2,4);
psd(h,s,'fs',fs),title('调制信号功率谱密度');
subplot(3,2,5);
plot(demodata1),title('解调输出');
axis([0 20000 - 2 2]);
subplot(3,2,6);
psd(h,demodata1,'fs',fs),title('解调输出功率谱密度');
%%%%%%%%%%%%%%%
% 通过高斯信道
figure(2)
subplot(2,2,1);
plot(s1),title('调制信号(Awgn)');
axis([0 500 - 5 5]);
subplot(2,2,2);
psd(h,s1,'fs',fs),title('调制信号功率谱密度(Awgn)');
subplot(2,2,3);
plot(s111),title('高斯噪声曲线');
axis([0 2000 - 5 5]);
subplot(2,2,4);
for i = 1:nb/2
plot(idata(i),qdata(i),'r + '),title('QPSK 信号星座图(Awgn)');hold on;
axis([- 2 2 - 2 2]);
plot(Awgn_ichsum(i),Awgn_qchsum(i),' * ');hold on;
legend('理论值(发射端)','实际值(接收端)');
end
%%%%%%%%%%%%%%%%%%
% 通过高斯信道再通过瑞利衰落信道
figure(3)
subplot(2,2,1)
plot(Ray_s),title('调制信号(Ray + Awgn)');
axis([0 500 - 5 5]);
subplot(2,2,2);
psd(h,Ray_s,'fs',fs),title('调制信号功率谱密度(Ray)');
subplot(2,2,3);
for i = 1:nb/2
plot(idata(i),qdata(i),'r + '),title('QPSK 信号星座图(Awgn + Ray)');hold on;
axis([- 2 2 - 2 2]);
```

```
plot(Ray_ichsum(i),Ray_qchsum(i),'*');hold on;
legend('理论值(发射端)','实际值(接收端)');
end
subplot(2,2,4)
semilogy(SNRindB2,theo_err_awgn_prb,'r'),title('误码率曲线');hold on;
semilogy(SNRindB1,smld_bit_awgn_err_prb,'r*');hold on;
semilogy(SNRindB2,theo_err_ray_prb);hold on;
semilogy(SNRindB1,smld_bit_ray_err_prb,'*');
xlabel('Eb/No');ylabel('BER');
legend('理论 AWGN','仿真 AWGN','理论 Rayleigh','仿真 Rayleigh');
```

（2）Create a new . model file as the sub function cm_sm32. m.

```
%%%%%%%%%%%%%%
the sub function cm_sm32.m
%%%%%%%%%%%%%%
% 文件 2
function [pb,ps] = cm_sm32(snr_in_dB)
%   [pb,ps] = cm_sm32(snr_in_dB)
%       CM_SM3 finds the probability of bit error and symbol error for
%       the given value of snr_in_dB, signal to noise ratio in dB.
N = 100;
E = 1;                          % energy per symbol
numofsymbolerror = 0;
numofbiterror = 0;
counter = 0;
snr = 10^(snr_in_dB/10);        % signal to noise ratio
sgma = sqrt(E/snr)/2;           % noise variance
s00 = [1 0]; s01 = [0 1]; s11 = [-1 0]; s10 = [0 -1];    % signal mapping
% generation of the data source
while(numofbiterror < 100)
for i = 1:N,
    temp = rand;                % a uniform random variable between 0 and 1
    if (temp < 0.25),           % with probability 1/4, source output is "00"
        dsource1(i) = 0; dsource2(i) = 0;
    elseif (temp < 0.5),        % with probability 1/4, source output is "01"
        dsource1(i) = 0; dsource2(i) = 1;
    elseif (temp < 0.75),       % with probability 1/4, source output is "10"
        dsource1(i) = 1; dsource2(i) = 0;
    else                        % with probability 1/4, source output is "11"
        dsource1(i) = 1; dsource2(i) = 1;
    end;
end;
% detection and the probability of error calculation
for i = 1:N,
    ray = raylrnd(0.8);
    n = sgma * randn(1,2);      % 2 normal distributed r.v with 0, variance sgma
    if ((dsource1(i) == 0) & (dsource2(i) == 0)),
        r = ray * s00 + n;
    elseif ((dsource1(i) == 0) & (dsource2(i) == 1)),
        r = ray * s01 + n;
```

```
            elseif ((dsource1(i) == 1) & (dsource2(i) == 0)),
                r = s10 * ray + n;
            else
                r = s11 * ray + n;
            end;
            % The correlation metrics are computed below
            c00 = dot(r,s00); c01 = dot(r,s01); c10 = dot(r,s10); c11 = dot(r,s11);
            % The decision on the ith symbol is made next
            c_max = max([c00,c01,c10,c11]);
            if (c00 == c_max), decis1 = 0; decis2 = 0;
            elseif (c01 == c_max), decis1 = 0; decis2 = 1;
            elseif (c10 == c_max), decis1 = 1; decis2 = 0;
            else decis1 = 1; decis2 = 1;
            end;
            % Increment the error counter, if the decision is not correct
            symbolerror = 0;
            if (decis1 ~= dsource1(i)),
                numofbiterror = numofbiterror + 1; symbolerror = 1;
            end;
            if (decis2 ~= dsource2(i)),
                numofbiterror = numofbiterror + 1; symbolerror = 1;
            end;
            if (symbolerror == 1),
                numofsymbolerror = numofsymbolerror + 1;
            end;
    end
    counter = counter + 1;
end
ps = numofsymbolerror/(N * counter);      % since there are totally N symbols
pb = numofbiterror/(2 * N * counter);      % since 2N bits are transmitted
```

(3) Create a new . model file as the sub function cm_sm33. m.

```
function [pb1,ps1] = cm_sm33(snr_in_dB)
% [pb,ps] = cm_sm32(snr_in_dB)
%      CM_SM3 finds the probability of bit error and symbol error for
%      the given value of snr_in_dB, signal to noise ratio in dB.
N = 100;
E = 1;                              % energy per symbol
snr = 10^(snr_in_dB/10);            % signal to noise ratio
sgma = sqrt(E/snr)/2;               % noise variance
s00 = [1 0]; s01 = [0 1]; s11 = [-1 0]; s10 = [0 -1]; % signal mapping
% generation of the data source
numofsymbolerror = 0;
numofbiterror = 0;
counter = 0;
while(numofbiterror < 100)
for i = 1:N,
    temp = rand;                    % a uniform random variable between 0 and 1
    if (temp < 0.25),               % with probability 1/4, source output is "00"
        dsource1(i) = 0; dsource2(i) = 0;
```

```
        elseif (temp < 0.5),                    % with probability 1/4, source output is "01"
            dsource1(i) = 0; dsource2(i) = 1;
        elseif (temp < 0.75),                   % with probability 1/4, source output is "10"
            dsource1(i) = 1; dsource2(i) = 0;
        else                                     % with probability 1/4, source output is "11"
            dsource1(i) = 1; dsource2(i) = 1;
        end;
    end;       % detection and the probability of error calculation


for i = 1:N,
    % the received signal at the detection, for the ith symbol, is:
    n = sgma * randn(1,2);                       % 2 normal distributed r.v with 0, variance sgma
    if ((dsource1(i) == 0) & (dsource2(i) == 0)),
        r = s00 + n;
    elseif ((dsource1(i) == 0) & (dsource2(i) == 1)),
        r = s01 + n;
    elseif ((dsource1(i) == 1) & (dsource2(i) == 0)),
        r = s10 + n;
    else
        r = s11 + n;
    end;
    % The correlation metrics are computed below
    c00 = dot(r,s00); c01 = dot(r,s01); c10 = dot(r,s10); c11 = dot(r,s11);
    % The decision on the ith symbol is made next
    c_max = max([c00,c01,c10,c11]);
    if (c00 == c_max), decis1 = 0; decis2 = 0;
    elseif (c01 == c_max), decis1 = 0; decis2 = 1;
    elseif (c10 == c_max), decis1 = 1; decis2 = 0;
    else decis1 = 1; decis2 = 1;
    end;
    % Increment the error counter, if the decision is not correct
    symbolerror = 0;
    if (decis1 ~= dsource1(i)), numofbiterror = numofbiterror + 1; symbolerror = 1;
    end;
    if (decis2 ~= dsource2(i)), numofbiterror = numofbiterror + 1; symbolerror = 1;
    end;
    if (symbolerror == 1), numofsymbolerror = numofsymbolerror + 1;
    end;
end
counter = counter + 1;
end
ps1 = numofsymbolerror/(N * counter);        % since there are totally N symbols
pb1 = numofbiterror/(2 * N * counter);       % since 2N bits are transmitted
```

Copy the results in the following blank（see Figure A. 40～Figure A. 42）.

3. The MATLAB realization of AMI & HDB₃

1) AMI

```
signal = [1 0 1 1 0 0 0 0 0 0 1 1 0 0 0 0 0 0 1 0] % code = AMI(s)
```

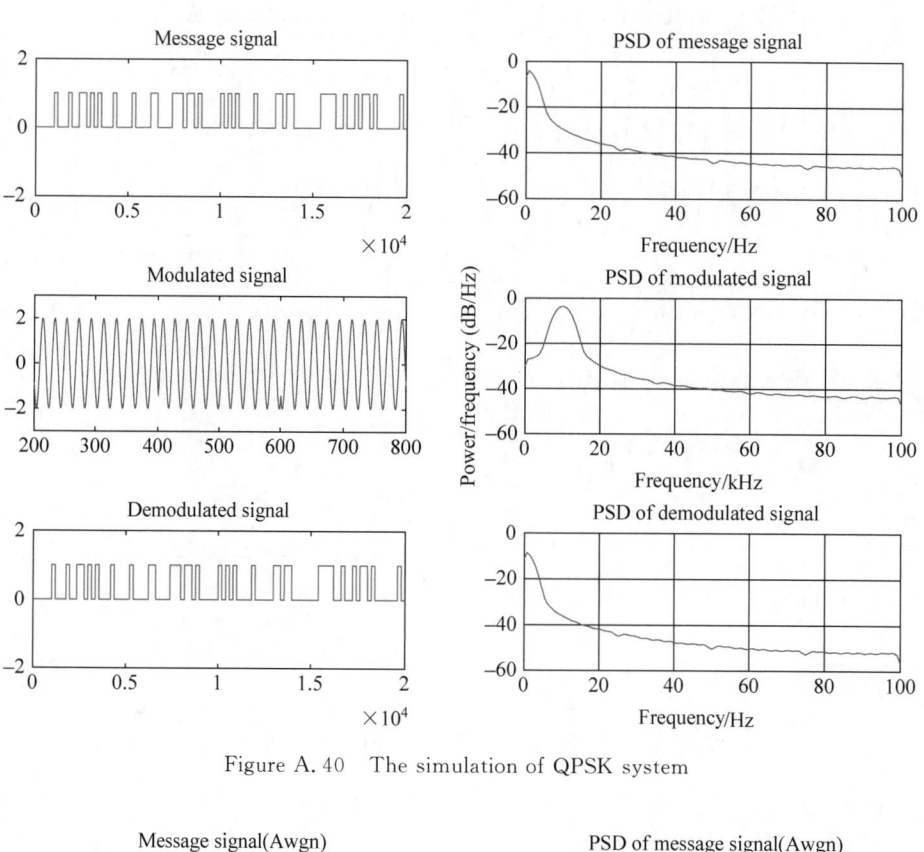

Figure A.40   The simulation of QPSK system

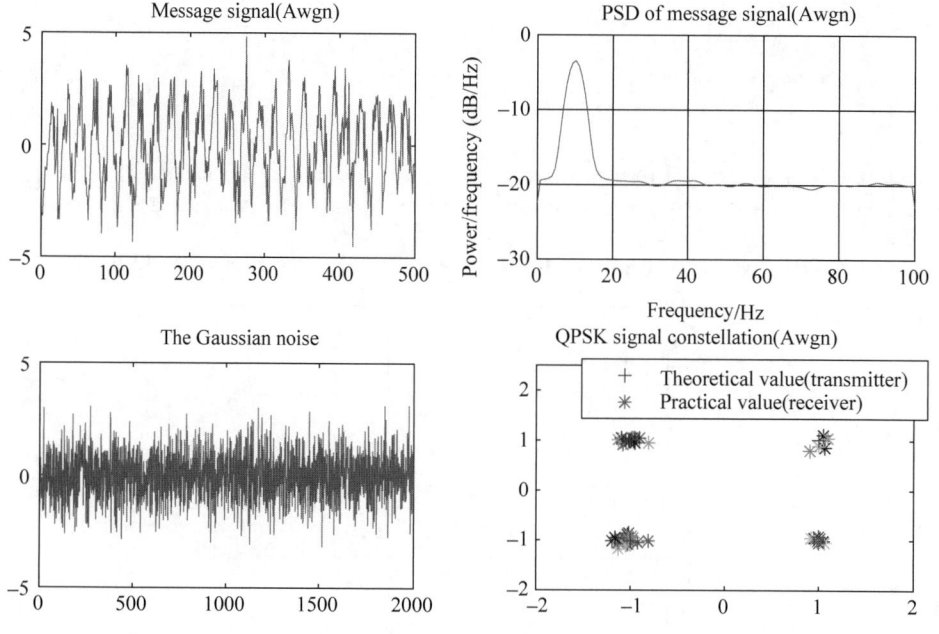

Figure A.41   The QPSK with Gaussian noise and its constellation diagram

```
status = - 1;
len = length(signal);
code = zeros(1,len);
```

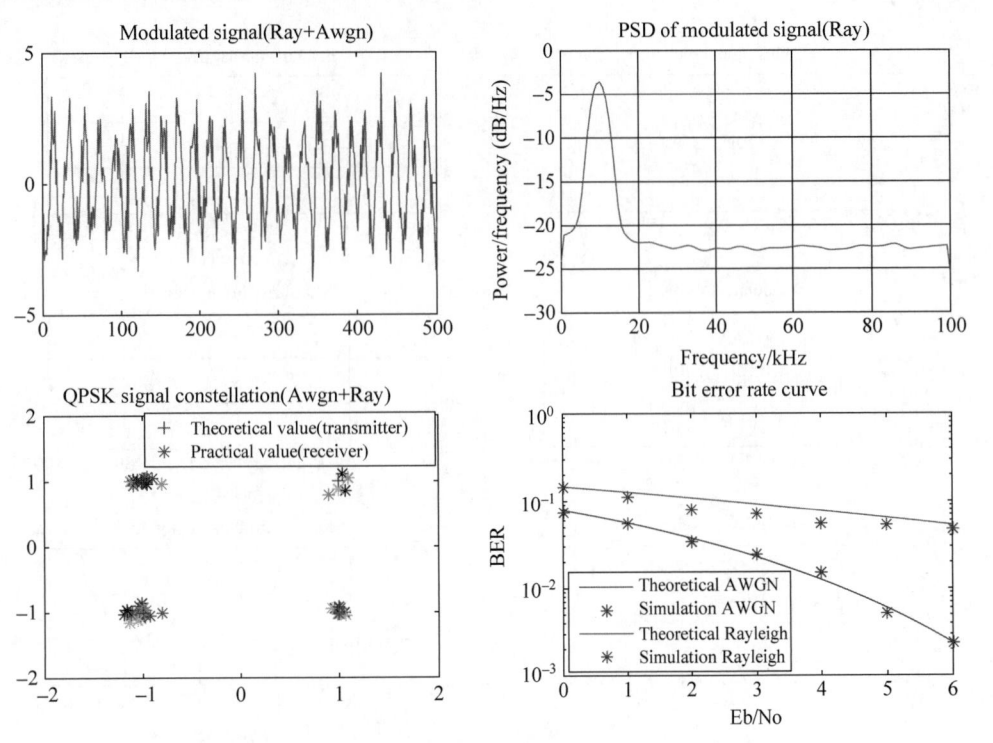

Figure A.42　The QPSK with Gaussian ＋Rayleigh noise

```
for i = 1:len
    if isequal(signal(i),1)
        code(i) = status;
        status = 0 - status;
    end
end
subplot(2,1,1);stairs([0:length(signal) - 1],signal);axis([0 length(signal) - 2 2]);
subplot(2,1,2);stairs([0:length(signal) - 1],code);axis([0 length(signal) - 2 2]);
```

Copy the HDB₃ encoder results in the following blanks（see Figure A.43）.

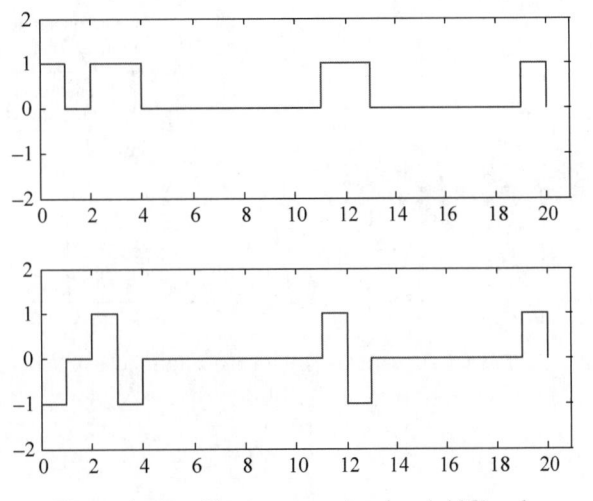

Figure A.43　The message signal and AMI code

2）HDB₃ 码

```
xn = [1 0 1 1 0 0 0 0 0 0 1 1 0 0 0 0 0 0 1 0];    % 输入单极性码
yn = xn; % 输出 yn 初始化
num = 0; % 计数器初始化
for k = 1:length(xn)
    if xn(k) == 1
        num = num + 1;                            % "1"计数器
            if num/2 == fix(num/2)                % 奇数个 1 时输出 - 1,进行极性交替
                yn(k) = 1;
            else
                yn(k) = - 1;
            end
    end
end

                                                  % HDB3 编码
num = 0;                                           % 连 0 计数器初始化
yh = yn;                                           % 输出初始化
sign = 0;                                          % 极性标志初始化为 0
V = zeros(1,length(yn));                           % V 脉冲位置记录变量
B = zeros(1,length(yn));                           % B 脉冲位置记录变量
for k = 1:length(yn)
    if yn(k) == 0
        num = num + 1;                            % 连"0"个数计数
        if num == 4                               % 如果 4 连"0"
            num = 0;                              % 计数器清零
            yh(k) = 1 * yh(k - 4);
                                                  % 让 0000 的最后一个 0 改变为与前一个非零符号相同极性的符号
            V(k) = yh(k);                         % V 脉冲位置记录
            if yh(k) == sign                      % 如果当前 V 符号与前一个 V 符号的极性相同
                yh(k) = - 1 * yh(k); % 则让当前 V 符号极性反转,以满足 V 符号间相互极性反转要求
                yh(k - 3) = yh(k);                % 添加 B 符号,与 V 符号同极性
                B(k - 3) = yh(k);                 % B 脉冲位置记录
                V(k) = yh(k);                     % V 脉冲位置记录
                yh(k + 1:length(yn)) = - 1 * yh(k + 1:length(yn));
                                                  % 并让后面的非零符号从 V 符号开始再交替变化
            end
            sign = yh(k);                         % 记录前一个 V 符号的极性
        end
    else
        num = 0;                                  % 当前输入为"1",则连"0"计数器清零
    end
end                                               % 编码完成
re = [xn',yn',yh',V',B'];                         % 结果输出: xn AMI HDB3 V&B 符号
                                % HDB3 解码
input = yh;                                        % HDB3 码输入
decode = input;                                    % 输出初始化
sign = 0;                                          % 极性标志初始化
for k = 1:length(yh)
    if input(k) ~= 0
        if sign == yh(k)                          % 如果当前码与前一个非零码的极性相同
            decode(k - 3:k) = [0 0 0 0];          % 则该码判为 V 码并将 * 00V 清零
        end
        sign = input(k);                          % 极性标志
```

```
        end
    end
    decode = abs(decode);                          % 整流
    error = sum([xn' - decode']);                  % 解码的正确性检验,作图
    subplot(3,1,1);stairs([0:length(xn) - 1],xn);axis([0 length(xn) - 2 2]);
    subplot(3,1,2);stairs([0:length(xn) - 1],yh);axis([0 length(xn) - 2 2]);
    subplot(3,1,3);stairs([0:length(xn) - 1],decode);axis([0 length(xn) - 2 2]);
    xn = [1 0 1 1 0 0 0 0 0 0 0 1 1 0 0 0 0 0 0 1 0];  % 输入单极性码
    yn = xn;                                        % 输出 yn 初始化
    num = 0;                                        % 计数器初始化
    for k = 1:length(xn)
        if xn(k) == 1
            num = num + 1;                          % "1"计数器
            if num/2 == fix(num/2)                  % 奇数个 1 时输出 - 1,进行极性交替
                yn(k) = 1;
            else
                yn(k) = - 1;
            end
        end
    end
```

Copy the HDB₃ encoder results in the following blanks(see Figure A. 44).

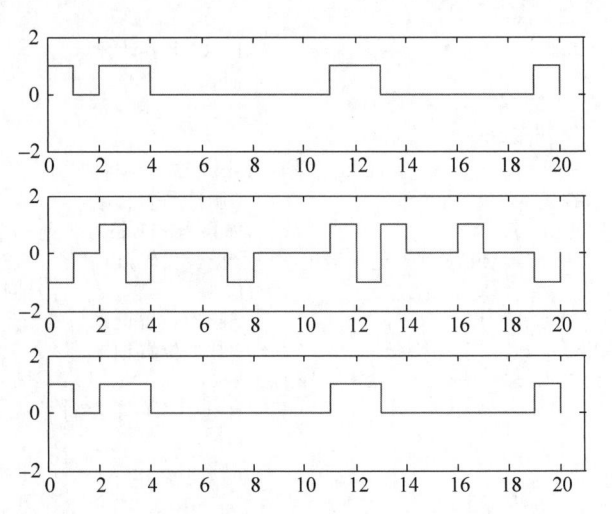

Figure A. 44   The message signal and HDB₃ code

4. Discussion the results

实验箱实验

本课程实验箱是由南京恒盾科技公司研制开发的综合性及验证性实验的组合,限于实验学时要求,只选取其中的 5 个实验进行学习,可以将其与 MATLAB 理论实验进行对比。其他实验可供学生课余选做。实验箱各模块分布如图 B.1 所示。

图 B.1　实验箱各模块示意图

实验一　各种模拟信号源实验

实验内容

(1) 测试各种模拟信号的波形。

(2) 测量信号音信号的波形。

1. 实验目的

(1) 熟悉各种模拟信号的产生方法及其用途。

（2）观察分析各种模拟信号波形的特点。

2. 电路工作原理

模拟信号源电路用来产生实验所需的各种音频信号：同步正弦波信号、非同步正弦波信号、语音信号和音乐信号等。

1) 同步信号源（同步正弦波发生器）

（1）功用。同步信号源用来产生与编码数字信号同步的 2kHz 正弦波信号，作为增量调制编码、PCM 编码实验的输入音频信号。在没有数字存储示波器的条件下，用它作为编码实验的输入信号，可在普通示波器上观察到稳定的编码数字信号波形。

（2）电路原理。图 B.2 为同步正弦信号发生器的电路图。它由 2kHz 方波信号产生器（图 B.2 中省略了）、高通滤波器、低通滤波器和输出电路 4 部分组成。

2kHz 方波信号由 CPLD 可编程器件内的逻辑电路通过编程产生。TP104 为其测量点。U107C 及周边的阻容网络组成一个截止频率为 ω_H 的二阶高通滤波器，用来滤除各次谐波。U107D 及周边的阻容网络组成一个截止频率为 ω_L 的二阶低通滤波器，用来滤除基波以下的杂波。两者组合成一个 2kHz 正弦波的带通滤波器，只输出一个 2kHz 正弦波，TP107 为其测量点。输出电路是由 BG102 和周边阻容元件组成的射极跟随器，起阻抗匹配、隔离与提高驱动能力的作用。

W104 用来改变高通滤波器反馈量的大小，使其工作在稳定的状态，W105 用来改变输出正弦波的幅度。

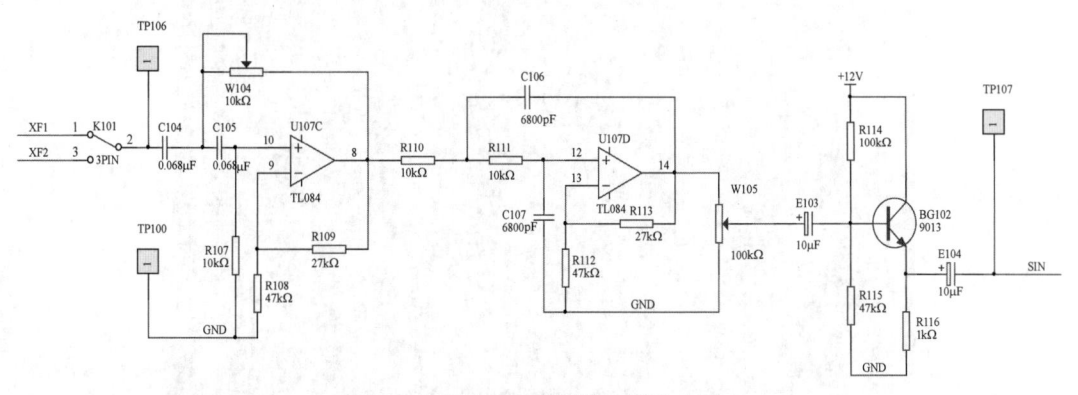

图 B.2　同步正弦信号发生器电路图

2) 话筒输入电路（麦克风电路）

（1）功用。话筒电路用来给驻极体话筒提供直流工作电压。

（2）工作原理。话筒电路如图 B.3 所示，V_{CC} 经分压器向话筒提供约 2.5V 工作电压，讲话时话筒与 R101 上的电压发生变化，其电压变化分量即为话音信号，经 E101 耦合输出，送往模拟信号输入选择电子开关。

3) 音乐信号产生电路

（1）功用。音乐信号产生电路用来产生音乐信号送往音频终端电路，以检查话音信道的开通情况及通话质量。

（2）工作原理。音乐信号产生电路见图 B.4。音乐信号由 U109 音乐片厚膜集成电路产生。该片的 1 脚为电源端，2 脚为控制端，3 脚为输出端，4 脚为公共地端。V_{CC} 经 R117、

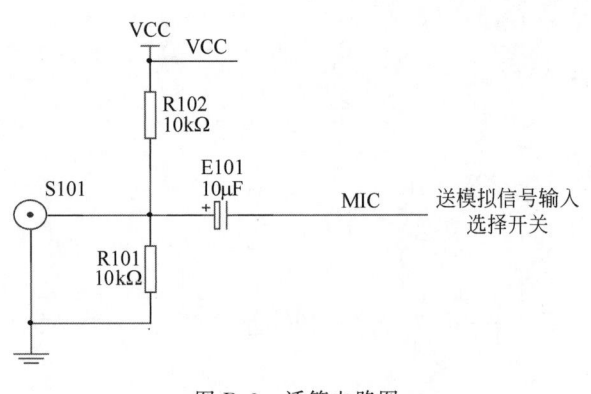

图 B.3 话筒电路图

D101 向 U109 的 1 脚提供 3.3V 电源电压,当 2 脚通过 K105 输入控制电压+3.3V 时,音乐片即有音乐信号从第 3 脚输出,经 E105 送往模拟信号输入选择电子开关。

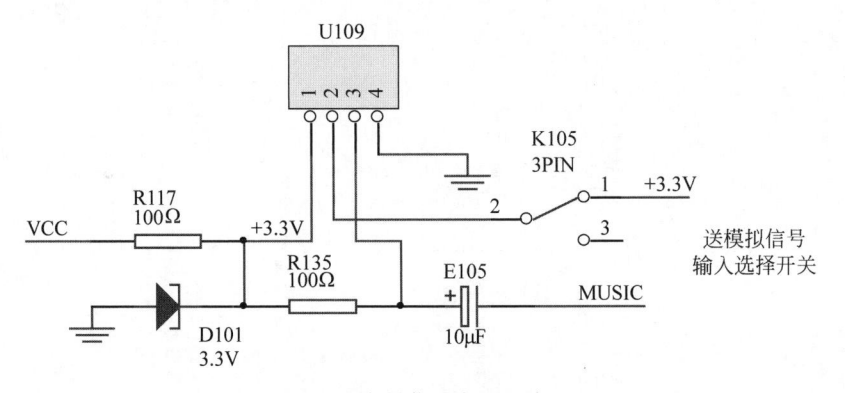

图 B.4 音乐信号产生电路图

4) 外加模拟信号输入电路

在一些特殊情况下,简易正弦波信号发生器不能满足实验要求,就要用外加信号源提供所需信号。例如要定量地测试通信话路的频率特性时需要使用频率与电平、输出阻抗都很稳定的频率范围很宽的音频测试信号,这就需要外接音频信号产生器或函数信号发生器。外加模拟信号输入电路为它们提供了连接到实验的接口电路。

5) 模拟电话输入电路

图 B.5 是用 PBL38710/1 电话集成电路组成的电话输入电路,J103 是手柄的送话器接口。讲话时话音信号从 TIPX 与 RINGX 引脚输入,经 U112 内部话音信号传输处理后从 VTX 与 RSN 引脚输出。输出信号分两路:一路经 K103 的 1-2 送往 PCM 编码器或经 K103 的 2-3 送往 PCM 编码器;另一路经 K104 的 1-2 或 2-3 送往话路终端接收滤波电路的 J105,选择后从话音信号输出电路的喇叭输出话音。

3. 实验内容

(1) 用示波器在相应测试点上测量各点波形:同步信号源、电话输入电路、话音输入电路、外加模拟信号输入电路。

(2) 熟悉上述各种信号的产生方法、来源及去处,了解信号流程。

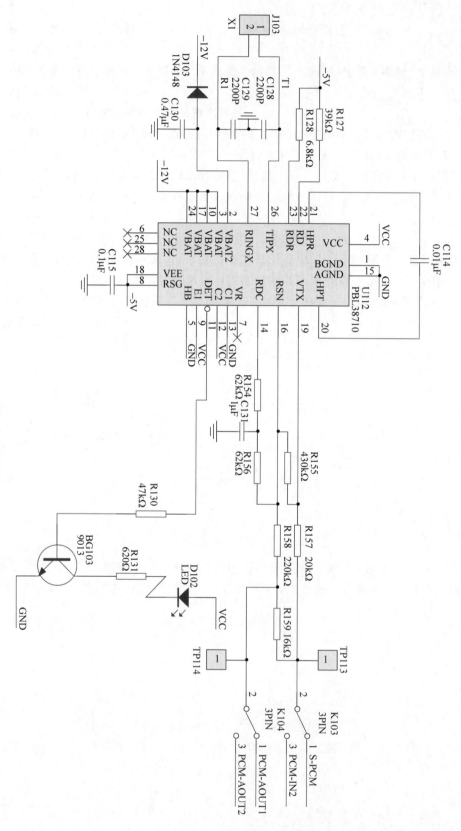

图 B.5　电话输入电路原理图

4. 实验步骤

(1) 用示波器测量 TP106、TP107、TP108、TP109、TP110、TP111、TP112、TP113、TP114 等各点波形。

(2) 测量音乐信号时用 K105 接通＋3.3V,此时 K105 短接 1-2,令音乐片加上控制信号,产生音乐信号输出。

5. 各测量点波形(以 PCM 为例)

TP106：由 CPLD(EPM7128SLC-15)分频产生的 2kHz 方波。

TP107：与工作时钟同步输出的 2kHz 正弦波信号。

TP108：0.3～3.4kHz 频率可调幅度可调的正弦波。

TP109：话路终端接收模拟信号输入。

TP110：音频功放输入信号。

TP111：音频输出信号。

TP112：话路终端发送模拟信号输出。

TP113：电话电路送往 PCM 编码器的话音信号。

TP114：电话电路送往话音终端接收滤波电路的话音信号。

实验二　脉冲编码调制 PCM

实验内容

(1) 用同步的简易信号观察 A 律 PCM 八比特编码的实验。

(2) 脉冲编码调制(PCM)及系统实验。

1. 实验目的

(1) 加深对 PCM 编码过程的理解。

(2) 熟悉 PCM 编码和译码专用集成芯片的功能和使用方法。

(3) 了解 PCM 系统的工作过程。

2. 实验电路工作原理

1) CM 的基本工作原理

脉冲调制就是把一个时间连续、取值连续的模拟信号变换成时间离散、取值离散的数字信号后在信道中传输。脉码调制就是对模拟信号先抽样,再对样值幅度量化、编码的过程。

所谓抽样,就是对模拟信号进行周期性扫描,把时间上连续的信号变成时间上离散的信号。该模拟信号经过抽样后还应当包含原信号中所有信息,也就是说,能无失真地恢复原模拟信号。它的抽样速率的下限是由抽样定理确定的。在该实验中,抽样速率采用 8kb/s。

所谓量化,就是将经过抽样得到的瞬时值进行幅度离散,即用一组规定的电平,把瞬时抽样值用最接近的电平值来表示。

一个模拟信号经过抽样量化后,得到已量化的脉冲幅度调制信号,它仅为有限个数值。

所谓编码,就是用一组二进制码组来表示每一个有固定电平的量化值。然而,实际上量化是在编码过程中同时完成的,故编码过程也称为模/数变换,记作 A/D。

由此可见,脉冲编码调制方式就是一种传递模拟信号的数字通信方式。

PCM 的原理如图 B.6 所示。话音信号先经防混叠低通滤波器进行脉冲抽样,变成

8kHz 重复频率的抽样信号（即离散的脉冲调幅 PAM 信号），然后将幅度连续的 PAM 信号用"四舍五入"法量化为有限个幅度取值的信号，再经编码，转换成二进制码。对于电话，CCITT 规定抽样率为 8kHz，每抽样值编 8 位码，即共有 $2^8 = 256$ 个量化值，因而每话路 PCM 编码后的标准数码率是 64kb/s。为解决均匀量化时小信号量化误差大、音质差的问题，在实际中采用不均匀选取量化间隔的非线性量化方法，即量化特性在小信号时分层密、量化间隔小，而在大信号时分层疏、量化间隔大，如图 B.7 所示。

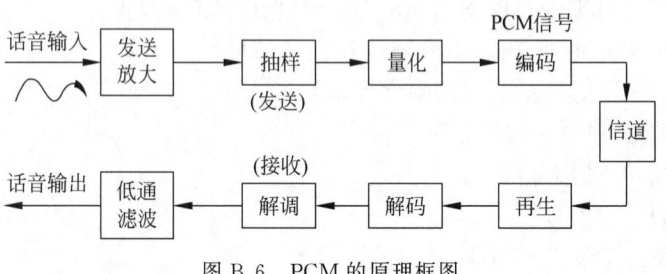

图 B.6 PCM 的原理框图

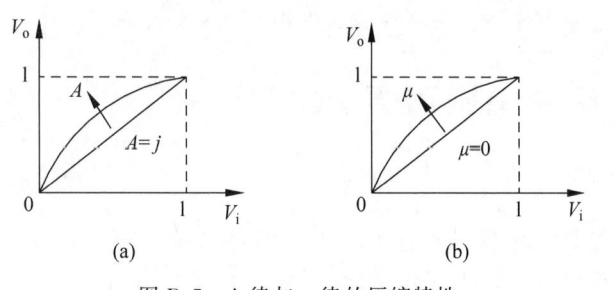

图 B.7 A 律与 μ 律的压缩特性

在实际中广泛使用的是两种对数形式的压缩特性：A 律和 μ 律。

A 律 PCM 用于欧洲和我国，μ 律用于北美和日本。它们的编码规律如图 B.8 所示。图中给出了信号抽样编码字与输入电压的关系，其中编码方式（1）为符号/幅度数据格式，Bit7 表示符号位，Bit6～0 表示幅度大小；（2）为 A 律压缩数据格式，它是（1）的 ADI（偶位反相）码；（3）为 μ 律压缩数据格式，它是由（1）的 Bit6～0 反相而得到，通常为避免 00000000 码出现，将其变成零抑制码 00000010。对压缩器而言，其输入输出归一化特性表示式为：

A 律

$$V_o = \begin{cases} \dfrac{AV_i}{1+\ln A} & 0 \leqslant V_i \leqslant \dfrac{1}{A} \\ \dfrac{1+\ln(AV_i)}{1+\ln A} & \dfrac{1}{A} \leqslant V_i \leqslant 1 \end{cases}$$

μ 律

$$V_o = \frac{1+\ln(1+\mu V_i)}{\ln(1+\mu)} \quad 0 \leqslant V_i \leqslant 1$$

2）PCM 编译码电路 TP3067 芯片介绍

（1）编译码器的简单介绍。

模拟信号经过编译码器时，在编码电路中，它要经过取样、量化、编码，如图 B.9(a)所

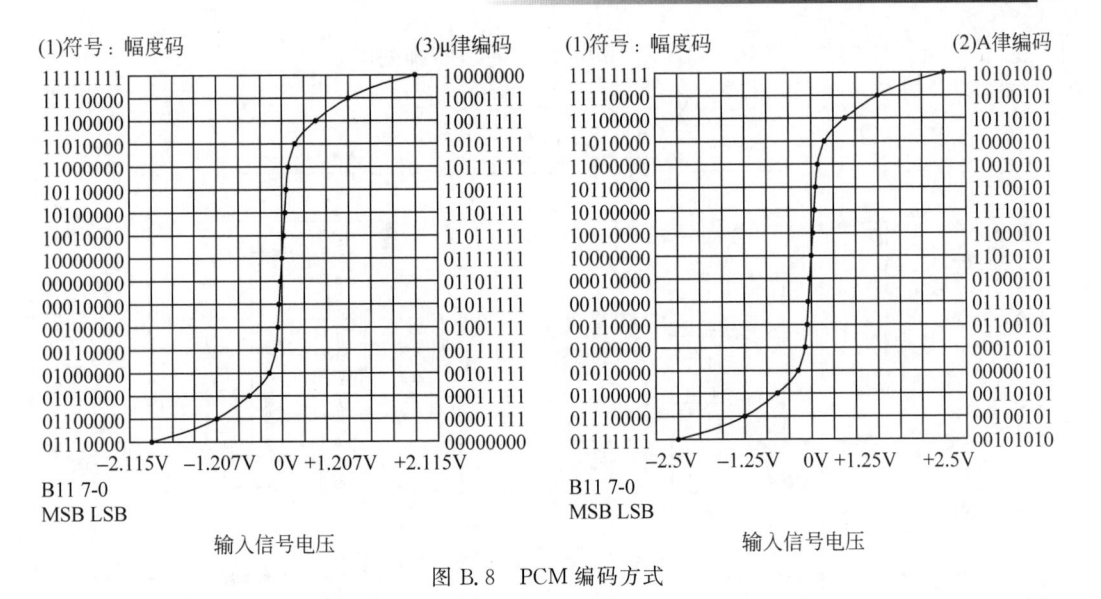

(1)符号：幅度码　　　　　　　　(3)μ律编码

```
11111111                              10000000
11110000                              10001111
11100000                              10011111
11010000                              10101111
11000000                              10111111
10110000                              11001111
10100000                              11101111
10010000                              11111111
10000000                              01111111
00000000                              01101111
00010000                              01011111
00100000                              01001111
00110000                              00111111
01000000                              00101111
01010000                              00011111
01100000                              00001111
01110000                              00000000
```
　　−2.115V　−1.207V　0V +1.207V　+2.115V

B11 7-0
MSB LSB

输入信号电压

(1)符号：幅度码　　　　　　　　(2)A律编码

```
11111111                              10101010
11110000                              10100101
11100000                              10110101
11010000                              10000101
11000000                              10010101
10110000                              11100101
10100000                              11110101
10010000                              11010101
10000000                              11010101
00010000                              01000101
00100000                              01110101
00110000                              01100101
01000000                              00010101
01010000                              00000101
01100000                              00110101
01110000                              00100101
01111111                              00101010
```
　　−2.5V　−1.25V　0V +1.25V　+2.5V

B11 7-0
MSB LSB

输入信号电压

图 B.8　PCM 编码方式

示。到底在什么时候被取样，在什么时序输出 PCM 码则由 A/D 控制来决定，同样 PCM 码被接收到译码电路后经过译码、低通滤波、放大，最后输出模拟信号，把这两部分集成在一个芯片上就是一个单路编译码器，它只能为一个用户服务，即在同一时刻只能为一个用户进行 A/D 及 D/A 变换。编码器把模拟信号变换成数字信号的规律一般有两种：一种是 μ 律十五折线变换法，它一般用在 PCM24 路系统中；另一种是 A 律十三折线非线性交换法，它一般应用于 PCM 30/32 路系统中，这是一种比较常用的变换法。模拟信号经取样后就进行 A 律十三折线变换，最后变成 8 位 PCM 码，在单路编译码器中，经变换后的 PCM 码是在一个时隙中被发送出去，这个时序号是由 A/D 控制电路来决定的，而在其他时隙时编码器是没有输出的，即对一个单路编译码器来说，它在一个 PCM 帧里只在一个由其 A/D 控制电路决定的时隙里输出 8 位 PCM 码，同样在一个 PCM 帧里，它的译码电路也只能在一个由它自己的 D/A 控制电路决定的时序里，从外部接收 8 位 PCM 码。其实单路编译码器的发送时序和接收时序还是可由外部电路来控制的，编译码器的发送时序由 A/D 控制电路来控制。定义为 FSx 和 FSr，要求 FSx 和 FSr 是周期性的，并且它的周期和 PCM 的周期要相同，都为 $125\mu s$，这样，每来一个 FSx，其 Codec 就输出一个 PCM 码，每来一个 FSr，其 Codec 就从外部输入一个 PCM 码。

图 B.9(b)是 PCM 的译码电路方框图，工作过程同图 B.9(a)相反，因此就不再讨论了。

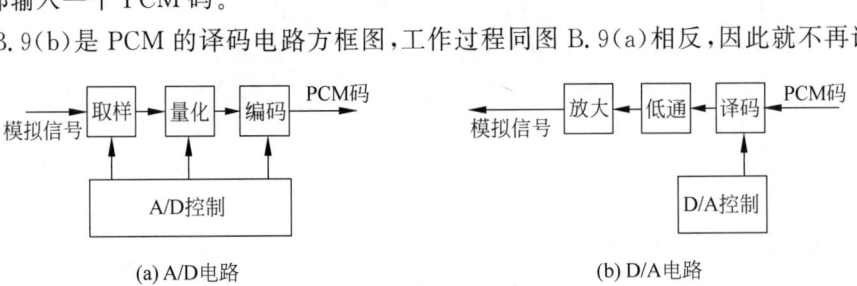

(a) A/D 电路　　　　　　　　　　(b) D/A 电路

图 B.9　A/D 及 D/A 电路框图

（2）本实验系统编译码器电路的设计。

我们所使用的编译码器是把编译码电路和各种滤波器集成在一个芯片上，它的框图如图 B.10 所示。该器件为 TP3067。图 B.11 是它的引脚排列图。

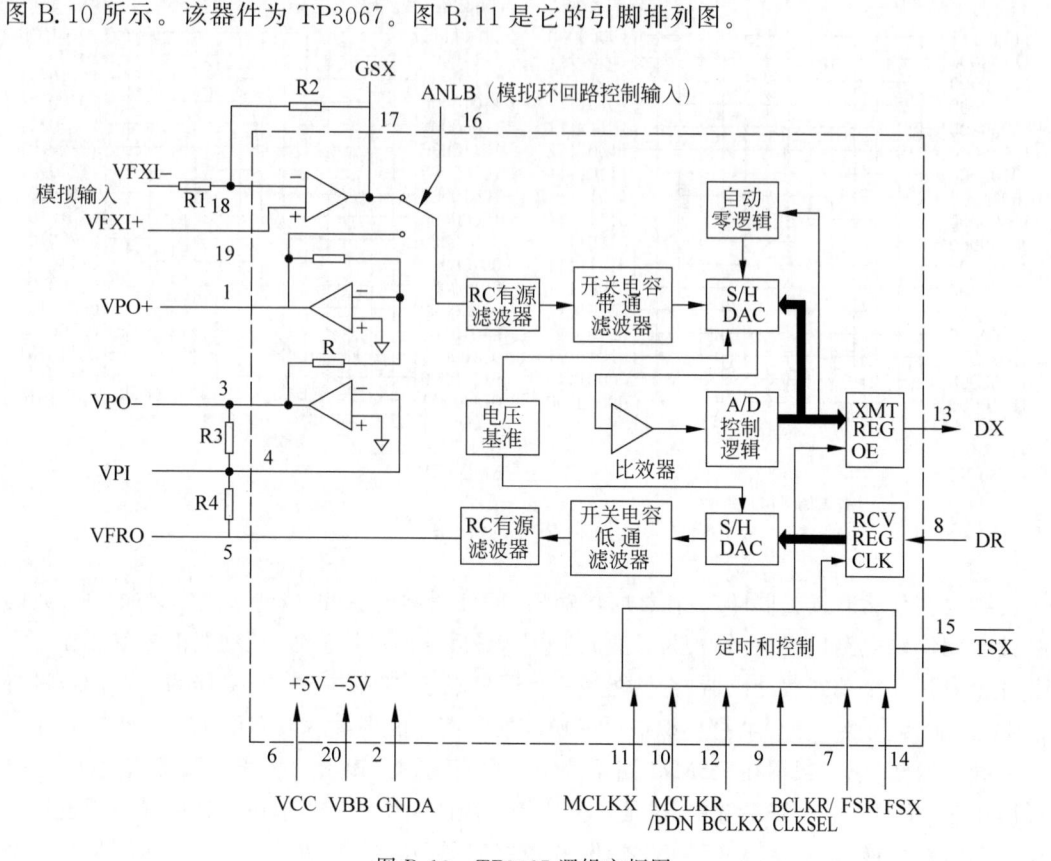

图 B.10　TP3067 逻辑方框图

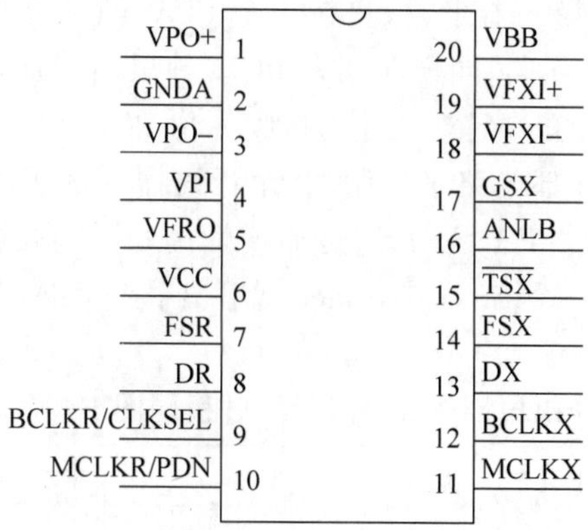

图 B.11　TP3067 引脚排列图

（3）引脚符号及其说明如下。

VPO+：接收功率放大器的同相输出。

GNDA：模拟地，所有信号均以该引脚为参考点。

VPO−：接收功率放大器的倒相输出。

VPI：接收功率放大器的倒相输入。

VFRO：接收滤波器的模拟输出。

VCC：正电源引脚，VCC＝＋5V±5%。

FSR：接收帧同步脉冲，FSR 为 8kHz 脉冲序列。

DR：接收帧数据输入。PCM 数据随着 FSR 前沿移入 DR。

BCLKR/CLKSEL：在 FSR 的前沿后把数据移入 DR 的位时钟，其频率可从 64k～2.48MHz。

MCLKR/PDN：接收主时钟，其频率可以为 1.536MHz、1.544MHz 或 2.048MHz。

MCLKX：发送主时钟，其频率可以是 1.536MHz、1.544MHz 或 2.048MHz。它允许与 MCLKR 异步，同步工作能实现最佳性能。

BCLKX：PCM 数据从 DX 上移出的位时钟，频率为 64k～2.048MHz，必须与 MCLKX 同步。

DX：由 FSX 启动的三态 PCM 数据输出。

FSX：发送帧同步脉冲输入，它启动 BCLKX 并使 DX 上 PCM 数据移到 DX 上。

ANLB：模拟环回路控制输入，在正常工作时必须置为逻辑 0，当拉到逻辑 1 时，发送滤波器和前置放大器输出被断开，改为和接收功率放大器的 VPO＋输出连接。

GSX：发送输入放大器的模拟输出。用来在外部调节增益。

VFXI－：发送输入放大器的倒相输入。

VFXI＋：发送输入放大器的非倒相输入。

VBB：负电源引脚，VBB＝－5V±5%。

（4）PCM 编译码电路。

PCM 编译码电路所需的工作时钟为 2.048MHz，FSR、FSX 的帧同步信号为 8kHz 窄脉冲，图 B.12 是短帧同步定时波形图，图 B.13 是时钟电路测量点波形图，图 B.14 是它的电原理图，图 B.15 是 PCM 编译码电路的波形图。

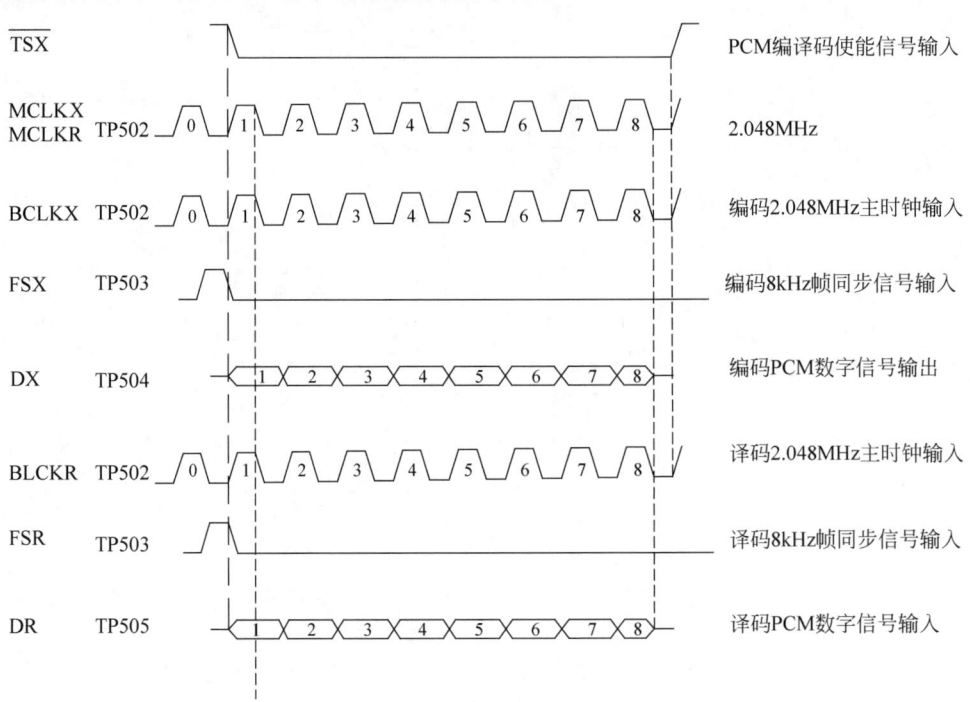

图 B.12 短帧同步定时

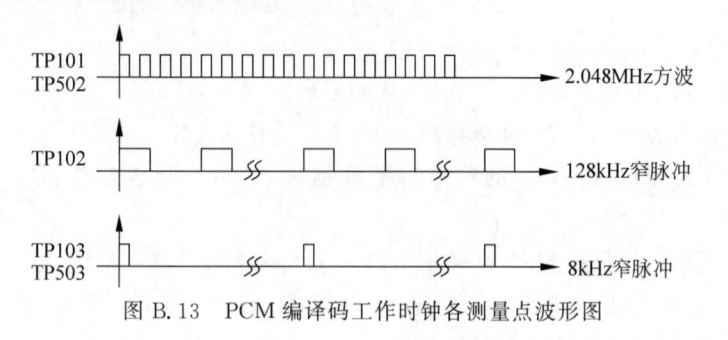

图 B.13　PCM 编译码工作时钟各测量点波形图

在本实验中选择 A 律变换，以 2.048Mb/s 的速率来传送信息，信息帧为无信令帧，它的发送时序与接收时序直接受 FSX 和 FSR 控制。还有一点，编译码器一般都有一个 PDN 降功耗控制端：PDN＝0 时，编译码能正常工作；PDN＝1 时，编译码器处于低功耗状态，这时编译码器的其他功能都不起作用。我们在设计时，可以实现对编译码器的降功耗控制。

3. 实验内容

(1) 用同步的简易信号观察 A 律 PCM 8 比特编码的实验。

(2) 脉冲编码调制（PCM）及系统实验。

4. 实验步骤及注意事项

(1) 在 PCM 系统送入两组信号，即

① 2048kHz 主时钟信号；

② 8kHz 收发分帧同步信号。

(2) 跳线开关放置：K501 的 2 脚和 3 脚、K502 的 1 脚和 2 脚、K503 的 1 脚和 2 脚。

(3) PCM 系统实验电路及参考波形图如图 B.14 及图 B.15 所示。

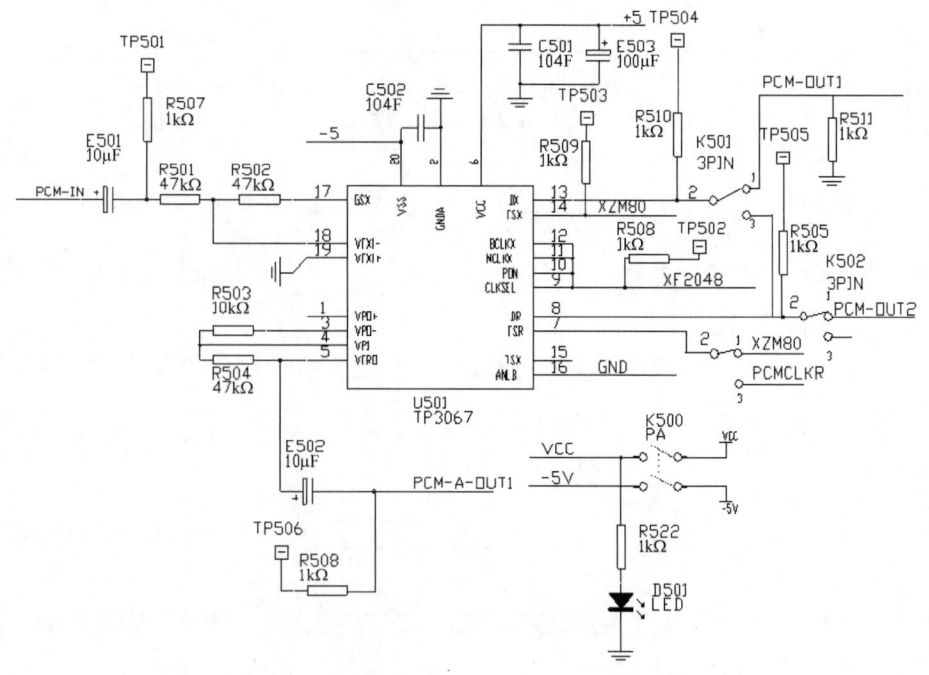

图 B.14　PCM 实验电路原理图

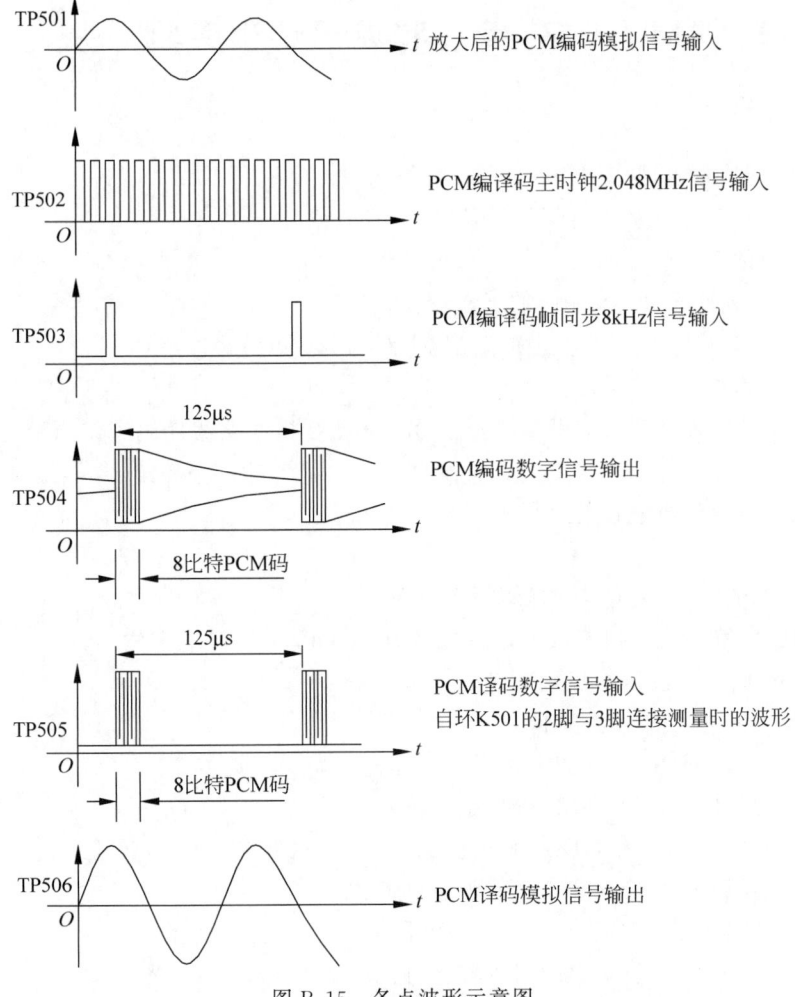

图 B.15 各点波形示意图

5. 测量点说明

TP501：输入信号由开关 J106 选择，若幅度过大，则被限幅电路限幅成方波，因此信号波形幅度尽量小一些，一般峰峰值为 1V 左右。调节幅度的方法是改变外部信号源的幅度大小，或调节电位器 W108。

TP502：频率为 2.048MHz 的主时钟信号，TP502 = TP101。

TP503：频率为 8kHz 的分帧同步信号，TP503 = TP103。

TP504：PCM 编码输出数字信号，数据的速率是 64kHz，为 8 比特编码，其中第一位为语音信号编码后的符号位，后 7 位为语音信号编码后的电平值。

TP505：PCM 译码输入数字信号，由开关 K501 的 2 脚与 3 脚相连，注意观察时示波器双通道，一定要和 TP503 帧同步信号同步观测。

TP506：PCM 译码输出模拟信号。

实验三 AMI/HDB$_3$ 编码和译码过程实验

实验内容

（1）AMI/HDB$_3$ 码型变换编码观察实验。

（2）AMI/HDB$_3$ 码型变换译码观察实验。

1. 实验目的

（1）熟悉 AMI/HDB$_3$ 编译码的工作过程。

（2）观察 AMI/HDB$_3$ 码型变换编译码电路的测量点波形。

2. 实验工作原理

在分析 HDB$_3$ 数字基带信号传输及 HDB$_3$ 码型变换线路编译码工作原理之前,首先对本实验电路中使用的 HDB$_3$ 专用集成电路 CD22103 芯片作一个介绍。

1）HDB$_3$ 专用集成电路 CD22103

（1）引脚功能说明。

第 1 脚:NRZ$_I$——发送端非归零码输入脚。

输入 HDB$_3$ 编码的非归零数据,它被编码时钟 CP$_1$ 的下降沿定位。

第 2 脚:CP$_1$——发送端编码时钟输入脚。

对 NRZ$_I$ 数据编码的输入时钟。

第 3 脚:AMI/HDB$_3$——码变换方式选择输入脚。

若 AMI/HDB$_3$ = L,为 NRZ-AMI 编译码;

若 AMI/HDB$_3$ = H,为 HDB$_3$ 编译码。

第 4 脚:NRZ$_O$——接收端非归零码输出脚。

译码后非归零数据,它定位于 CP$_2$ 上升沿。

第 5 脚:CP$_2$——接收端解码时钟输入脚。

对 AIN、BIN——数据进行解码的时钟信号。

第 6 脚:SET——输入 HDB$_3$ 码连零告警置位端。

第 7 脚:AIS——HDB$_3$ 码连零告警输出端。当 $\overline{SET}$ = L 时,译码计数器清零,此后若 AIS=L,表示前段在 $\overline{SET}$ = H 期间译码过程中出现不少于 3 个 0;若 AIS=H,表示出现少于 3 个 0。当 SET = H 时,使译码计数器工作,进行连 0 统计。

第 8 脚:GND——地。

第 9 脚:ERR——接收端误码检测输出端,它以违反 HDB$_3$ 编码规律为标准,统计接收 HDB$_3$ 码的错误情况。若 HDB$_3$ 码出现同极性的 3 个 1 时,则 ERR=H。

第 10 脚:CP$_3$——接收端时钟输出端,提供为位同步需要的时钟信息,若 $\overline{LTE}$ =L, CP$_3$=AIN+BIN;若 $\overline{LTE}$ =H,则 CP$_3$=OUT$_1$+OUT$_2$。

第 11 脚:AIN——解码输入端（+）。

第 12 脚:$\overline{LTE}$——工作自环控制输入脚,自环/工作控制信号,当 $\overline{LTE}$=L 时,为正常工作状态,编解码器独立,以异步方式工作;当 $\overline{LTE}$=H 时,内部将 OUT$_1$ 与 AIN,OUT$_2$ 与 BIN 短接,CP$_3$=OUT$_1$+OUT$_2$,电路处于环路测试状态,此时 NRZ 相对于 NRZ$_0$ 延时

6.5 个时钟周期。

第 13 脚: BIN——解码输入端(一),表示接收的欲解码两路单极性 HDB_3(+)、(一)码序列,它输入后被解码时钟 CP_2 的上升沿抽样。

第 14 脚: OUT_1——发送端编码输出端(一)。

第 15 脚: OUT_2——发送端编码输出端(+),表示编码后 HDB_3 的两路单极性码序列,通常经变压器合成三电平 HDB_3 码。HDB_3 码输出。

第 16 脚: V_+——正电源,电压通常为 $+5V\pm0.25V$。

(2) 集成电路 CD22103 功能框图如图 B.16 所示。

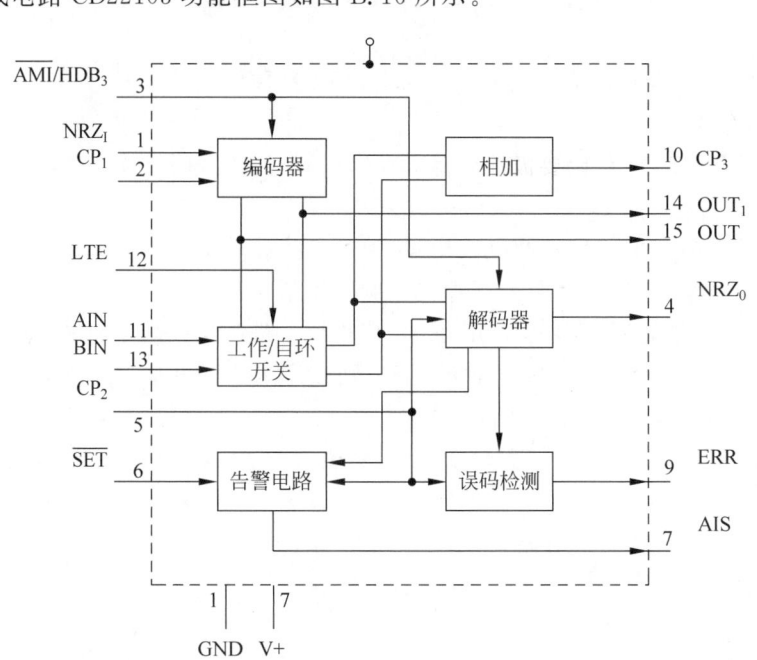

图 B.16 集成电路 CD22103 功能框图

2) HDB_3 电路的工作原理

AMI 码的全称是传号交替反转码。这是一种将消息代码 0(空号)和 1(传号)按如下规则进行编码的码: 代码的 0 仍变换为传输码的 0,而把代码中的 1 交替地变换为传输码的 $+1$、-1、$+1$、-1……由于 AMI 码的信号交替反转,故由它决定的基带信号将出现正负脉冲交替,而 0 电位保持不变的规律。由此看出,这种基带信号无直流成分,且只有很小的低频成分,因而它特别适宜在不允许这些成分通过的信道中传输。

从 AMI 码的编码规则看出,它已从一个二进制符号序列变成了一个三进制符号序列,而且也是一个二进制符号变换成一个三进制符号。把一个二进制符号变换成一个三进制符号所构成的码称为 1B/1T 码型。

AMI 码除具有上述特点外,还有编译码电路简单及便于观察误码情况等优点,它是一种基本的线路码,并得到广泛采用。但是,AMI 码有一个重要缺点,即当它用来获取定时信息时,由于它可能出现长的连 0 串,因而会造成提取定时信号的困难。

为了保持 AMI 码的优点而克服其缺点,人们提出了许多种类的改进 AMI 码,HDB_3 码就是其中有代表性的码。

HDB_3 码是三阶高密度码的简称。HDB_3 码保留了 AMI 码所有的优点（如前所述），还可将连码限制在 3 个以内，克服了 AMI 码如果连 0 过多对提取定时时钟不利的缺点。HDB_3 码的功率谱与 AMI 码类似。由于 HDB_3 码的诸多优点，所以 CCITT 建议将 HDB_3 码作为 PCM 传输系统的线路码型。

如何由二进制码转换成 HDB_3 码呢？

HDB_3 码的编码规则如下：

（1）二进制序列中的 0 码在 HDB_3 码中仍编为 0 码，但当出现 4 个连 0 码时，用取代节 000V 或 B00V 代替。取代节中 V 码、B 码均代表 1 码，它们可正可负（即 $V_+ = +1$，$V_- = -1$，$B_+ = +1$，$B_- = -1$）。

（2）取代节的安排顺序是：先用 000V，当它不能用时，再用 B00V，000V 取代节的安排要满足以下两个要求：

- 各取代节之间的 V 码要极性交替出现（为了保证传号码极性交替出现，不引入直流成分）。
- V 码要与前一个传号码的极性相同（为了在接收端能识别出哪个是原始传号码，哪个是 V 码和 B 码，从而恢复成原二进制码序列）。

当上述两个要求能同时满足时，用 000V 代替原二进制码序列中的 4 个 0（用 $000V_+$ 或 $000V_-$）；而当上述两个要求不能同时满足时，则改用 B00V（B_+00V_+ 或 B_-00V_-，实质上是将取代节 000V 中第一个 0 码改成 B 码）。

（3）HDB_3 码序列中的传号码（包括 1 码、V 码和 B 码）除 V 码外要满足极性交替出现的原则。

下面举例说明，如何将二进制码转换成 HDB_3 码。

二进制码序列：

1 0 0 0 0 1 0 1 0 0 0 0 0 1 1 1 0 0 0 0 0 0 0 0 0 1

HDB_3 码序列：

$V_+ -1 0 0 0 V_- +1 0 -1 B_+ 0 0 V 0 -1 +1 -1 0 0 0 V_-$ $B_+ 0 0 V_+ 0 -1$

从上例可以看出两点：

- 当两个取代节之间原始传号码的个数为奇数时，后边取代节用 000V；当两个取代节之间原始传号码的个数为偶数时，后边取代节用 B00V。
- V 码破坏了传号码极性交替出现的原则，所以叫破坏点；而 B 码未破坏传号码极性交替出现的原则，叫非破坏点。

虽然 HDB_3 码的编码规则比较复杂，但译码却比较简单。从上述原理可以看出，每一个破坏符号 V 总是与前一非 0 符号同极性（包括 B 在内）。也就是说，从收到的符号序列中可以容易地找到破坏点 V，所以可以断定 V 符号及其前面的 3 个符号必是连 0 符号，从而恢复 4 个连 0 码，再将所有 -1 变成 $+1$ 后便得到原消息代码。编码工作波形如图 B.17 所示。

3）电路的工作过程

译码是编码的逆过程。其波形如图 B.18 所示。但 CP_2 应比译码输入（AIN、BIN）稍有延时。环路测试由 LTE 控制，若 LTE=H，则 OUT_1、OUT_2 内部短接到对应的 AIN、BIN，此时 NRZ_0 应为 NRZ_i，但延后 8 个时钟周期左右。CP_3 为 AIN、BIN 相加波形，供接收端提取时钟用。

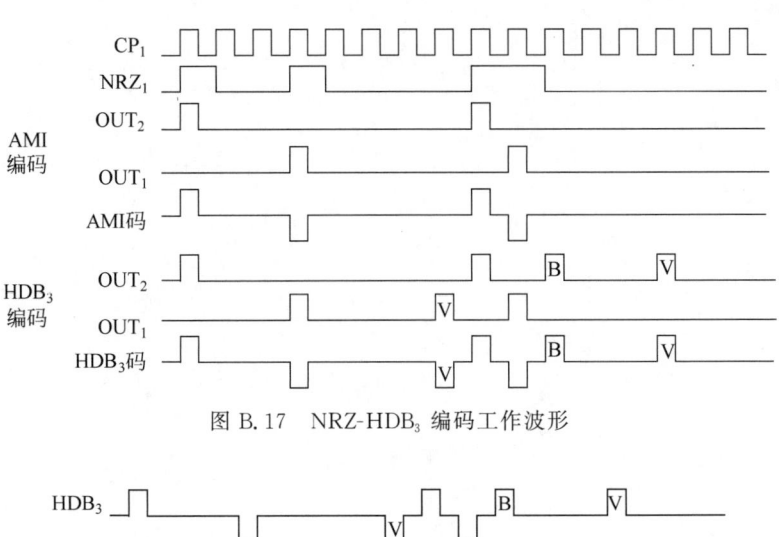

图 B.17   NRZ-HDB₃ 编码工作波形

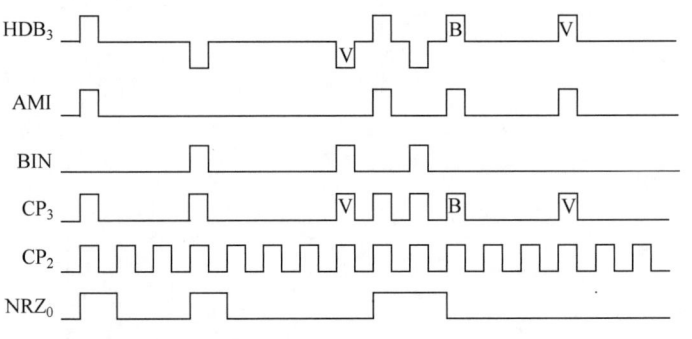

图 B.18   HDB₃ 译码工作波形

4) 实验电路工作原理

在实验系统中,电路原理图如图 B.19 所示。采用了 UA01(SC22103 专用芯片)实现 AMI/HDB₃ 的编译码实验,在该电路模块中,没有采用复杂的线圈耦合的方法来实现 HDB₃ 码字的调试,而是采用 UA02A(TL084)对 HDB₃ 的输出进行变换。

输入的码流由 UA01 的 1 脚在 2 脚时钟信号的推动下输入,HDB₃ 与 AMI 由 KA01 选择。编码之后的结果在 UA01 的 14 脚和 15 脚输出。而后在电路上直接由 UA01 的 11 脚和 13 脚返回,再由 UA03 进行译码。正确译码之后,TPA01 与 TPA08 的波形应一致,但由于 HDB₃ 的编译码规则较复杂,当前输出的 HDB₃ 码字可能与前 4 个码字有关,因而 HDB₃ 的编译码延时较大。

AMI 与 HDB₃ 的选择可通过 KA01 设置,当 KA01 设置在 1-2 状态时,UA01 完成 HDB₃ 编译码过程;当 KA01 设置在 1-2 状态时,UA01 完成 AMI 编译码过程。AMI/HDB₃ 的编译码工作波形如图 B.20 所示(为了便于说明,编码电路各波形的延时都已略去)。

3. 实验内容

(1) AMI/HDB₃ 码型变换编码观察实验。

(2) AMI/HDB₃ 码型变换译码观察实验。

4. 实验步骤

(1) 按下按键开关:K01、K02、KA00。

(2) 跳线开关设置:KA01 置 1-2(AMI)或 2-3(HDB₃)、KA02 置 2-3、KA03 置 2-3。

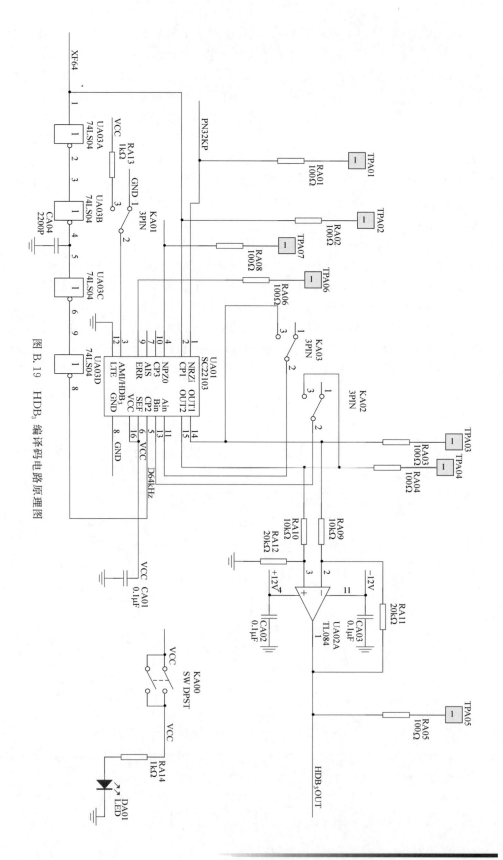

图 B.19 HDB₃ 编译码电路原理图

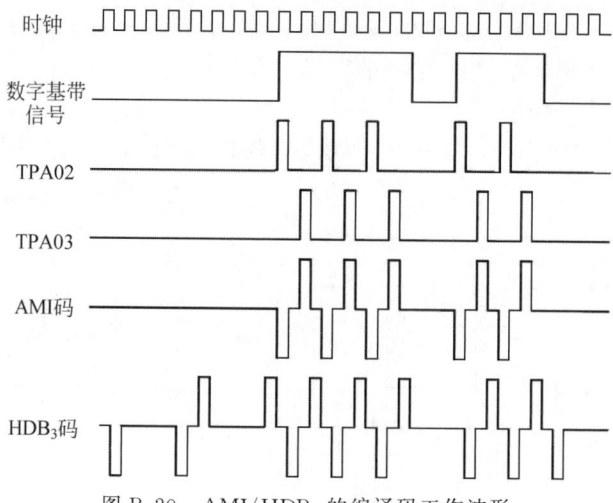

图 B.20 AMI/HDB₃ 的编译码工作波形

5. 测量点说明

TPA01：发送端数字基带信码输入，码型为：000011101100101，如反相属正常，不影响正常实验。

TPA02：发送端 64kHz HDB₃ 编码的工作时钟输入。

TPA03：AMI/HDB₃ 编码时 OUT1 的输出波形。

TPA04：AMI/HDB₃ 编码时 OUT2 的输出波形。

TPA05：AMI/HDB₃ 编码输出波形。

TPA06：正常工作时为低电平。

TPA07：接收端译码数字基带信码输出，码型同 TPA01，波形有延时。

实验四 FSK 调制解调实验

实验内容

（1）频率键控（FSK）调制实验。

（2）频率键控（FSK）解调实验。

1. 实验目的

（1）理解 FSK 调制的工作原理及电路组成。

（2）理解利用锁相环解调 FSK 的原理和实现方法。

2. 实验电路工作原理

数字频率调制是数据通信中使用较早的一种通信方式。由于这种调制解调方式容易实现，抗噪声和抗衰减性能较强，因此在中低速数据传输通信系统中得到了较为广泛的应用。

数字调频又可称作移频键控 FSK，它是利用载频频率变化来传递数字信息的。数字调频信号可以分为相位离散和相位连续两种情形。若两个振荡频率分别由不同的独立振荡器提供，其相位互不相关，这就叫相位离散的数字调频信号；若两个振荡频率由同一振荡信号源提供，只是对其中一个载频进行分频，这样产生的两个载频就是相位连续的数字调频信号。在本实验电路中,由实验一提供的载频频率经过本实验电路分频而得到的两个不同频率的载频信号,则为相位连续的数字调频信号。

1) FSK 调制电路工作原理

FSK 调制解调原理框图,如图 B.21 所示,图 B.22 是其调制电路原理图。

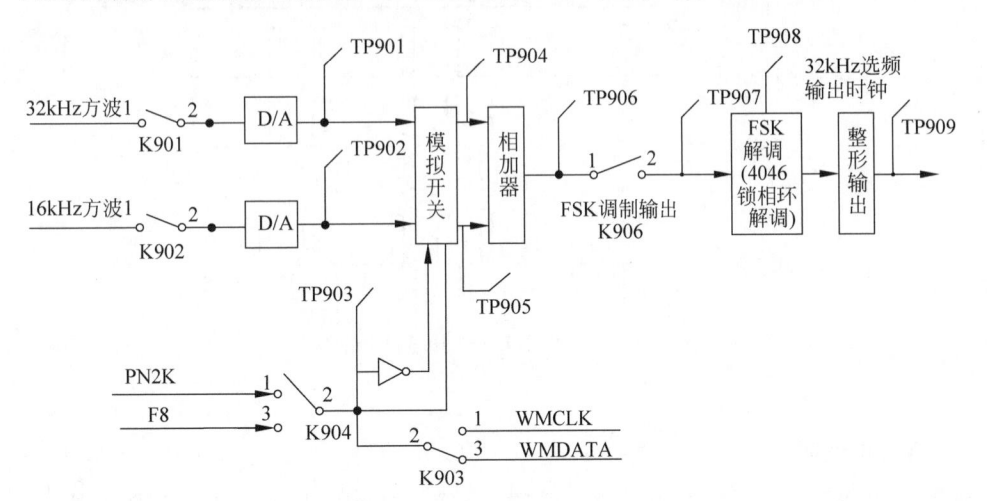

图 B.21　FSK 调制解调原理框图

输入的基带信号由转换开关 K904 转接后分成两路:一路控制 $f_1=32\text{kHz}$ 的载频,另一路经倒相去控制 $f_2=16\text{kHz}$ 的载频。当基带信号为 1 时,模拟开关 1 打开,模拟开关 2 关闭,此时输出 $f_1=32\text{kHz}$;当基带信号为 0 时,模拟开关 1 关闭,模拟开关 2 开通。此时输出 $f_2=16\text{kHz}$,于是可在输出端得到已调的 FSK 信号。

电路中的两路载频(f_1、f_2)由内时钟信号发生器产生,经过开关 K901 和 K902 送入。两路载频分别经射极跟随器、选频滤波、射极跟随器,再送至模拟开关 U901:A 与 U901:B(4066)。

2) FSK 解调电路工作原理

FSK 集成电路模拟锁相环解调器由于性能优越、价格低廉、体积小,得到了越来越广泛的应用。解调电路原理图如图 B.23 所示。

FSK 集成电路模拟锁相环解调器的工作原理是十分简单的,只要在设计锁相环时,使它锁定在 FSK 的一个载频 f_1 上,对应输出高电平,而对另一载频 f_2 失锁,对应输出低电平,那么在锁相环路滤波器输出端就可以得到解调的基带信号序列。

FSK 锁相环解调器中的集成锁相环选用 MC14046。

压控振荡器的中心频率设计为 32kHz。图 B.23 中 R924、R925、CA901 主要用来确定压控振荡器的振荡频率。R929、C904 构成外接低通滤波器,其参数选择要满足环路性能指标的要求。从要求环路能快速捕捉、迅速锁定来看,低通滤波器的通带要宽些;从提高环路的跟踪特性来看,低通滤波器的通带又要窄些。因此电路设计应在满足捕捉时间的前提下,尽量减小环路低通滤波器的带宽。

当输入信号为 16kHz 时,环路失锁。此时环路对 16kHz 载频的跟踪破坏。

可见,环路对 32kHz 载频锁定时输出高电平,对 16kHz 载频失锁时就输出低电平。只要适当选择环路参数,使它对 32kHz 锁定,对 16kHz 失锁,则在解调器输出端就得到解调输出的基带信号序列。关于 FSK 调制原理波形如图 B.24 所示。

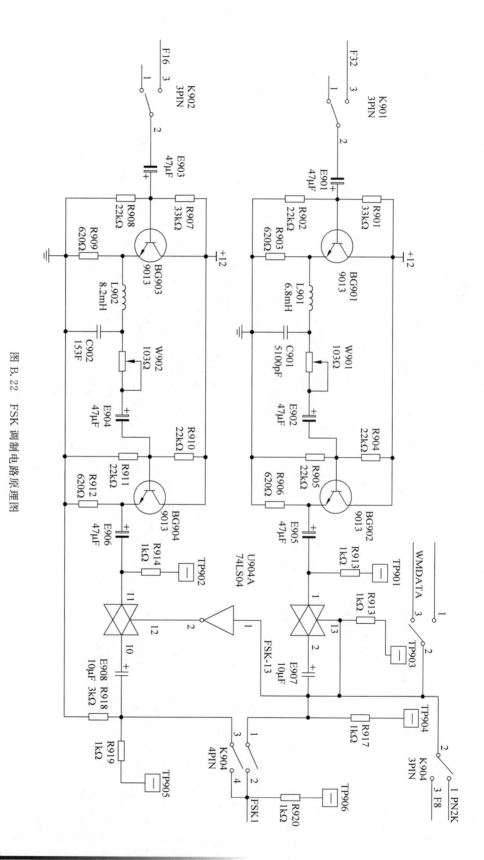

图 B.22　FSK 调制电路原理图

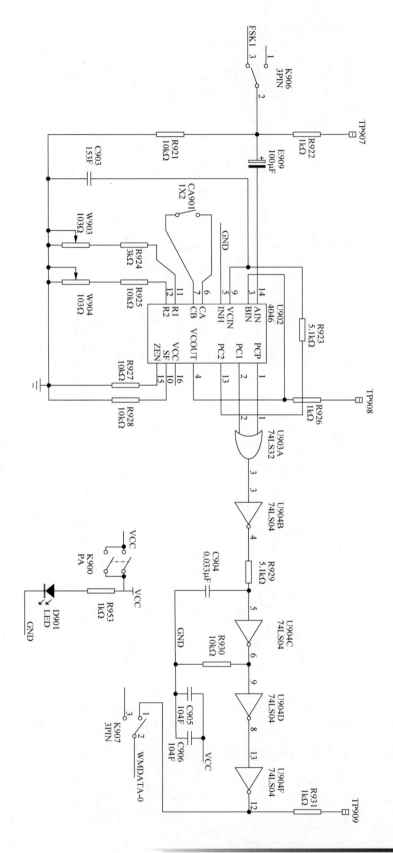

图 B.23 FSK 解调电路原理图

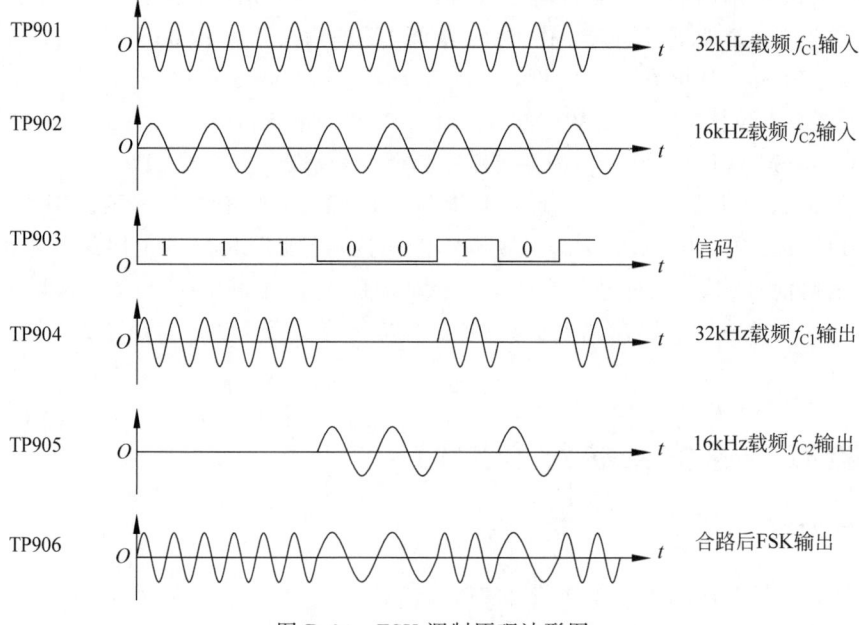

图 B.24　FSK 调制原理波形图

3. 实验内容

测试 FSK 调制解调电路 TP901～TP909 各测量点波形,并作详细分析。

(1) 按下按键开关:K01、K02、K900。

(2) 跳线开关设置:K901$_{2\text{-}3}$、K902$_{2\text{-}3}$。

K904$_{1\text{-}2}$、2kHz 的伪随机码,码序列为:000011101100101。

K904$_{2\text{-}3}$、8kHz 方波。

做 FSK 解调实验时,K904$_{1\text{-}2}$、K903$_{1\text{-}2}$。

(3) 在 CA901 插上电容,使压控振荡器工作在 32kHz,电容为 1800～2400pF。

(4) 注意选择不同的数字基带信号的速率。有 1110010 码(2kHz)、1010 交替码 (8kHz)。由信号转接开关 K904 进行选择。

(5) 接通开关 K906 的 2 脚和 3 脚,输入 FSK 信号给解调电路,注意观察 1、0 码内所含载波的数目。

(6) 观察 FSK 解调输出 TP907～TP909 波形,并作记录,并同时观察 FSK 调制端的基带信号,比较两者波形,观察是否有失真。

4. 测量点说明

TP901:32kHz 载频信号,由 K901 的 2 脚与 3 脚相连,可调节电位器 W901 改变幅度。

TP902:16kHz 载频信号,由 K902 的 2 脚与 3 脚相连,可调节电位器 W902 改变幅度。

TP903:作为 $F=2$kHz 或 8kHz 的数字基带信码信号输入,由开关 K904 决定。K904 的 1 脚与 2 脚相连:码元速率为 2kHz 的 000011101100101 码;K904 的 2 脚与 3 脚相连:码元速率为 8kHz 的 10101010 码。

TP904:32kHz 基带 FSK 调制信号输出。

TP905:16kHz 基带 FSK 调制信号输出。

TP906：FSK 调制信号叠加后输出，送到 FSK 解调电路的由输入开关 K905 控制。

TP907：FSK 解调信号输入。由 FSK 解调电路的输入开关 K906 的 2 脚与 3 脚接入。

TP908：FSK 解调电路工作时钟，正常工作时应为 32kHz 左右，频偏不大于 2kHz，若有偏差，可调节电位器 W903 或 W904 和改变 CA901 的电容值。

TP909：FSK 解调信号输出，即数字基带信码信号输出，波形同 TP905。

注：在 FSK 解调时，K904 只能是 1 脚与 2 脚相连，即解调出码元速率为 2kHz 的 000011101100101 码。K904 的 2 脚与 3 脚不能相连，否则 FSK 解调电路解调不出此时的数字基带信码信号，因为此时 $F=8\text{kHz}$，$f_{c2}=16\text{kHz}$，所以不满足 $4F\leqslant f_{c1}$ 的关系，因为此时它们的频谱重叠了。所以在此项实验做完后，应注意把开关 K904 设置成 1 脚与 2 脚相连接的位置上。

实验五　通信系统综合实验

实验内容

(1) 语音信号及通信系统原理综合实验。

(2) 用单台实验箱实现单工通信系统实验。

(3) 用两台实验箱实现双工通信系统实验。

1. 实验目的

(1) 熟悉数字通信系统各级信号的波形。

(2) 理解信号在信道传输过程中的变换原理和方法。

(3) 了解数字通信系统性能的测试方法。

2. 实验电路工作原理

为了使学生对数字通信系统有一个感性认识，进一步加深对通信系统工作的理解，本实验在前面实验的基础上，有代表性地选择了增量编码调制实验、二相 PSK（DPSK）调制解调系统实验、信号发生器系统实验和中央集中控制器系统实验的基础上，组成一个较完整的通信系统综合实验，也即模拟一个简易数字通信系统。图 B.25 是该系统较详细的电路方框图。下面就以图 B.25 为例介绍各部分的功能。

方框 ①：音频信号发生器。

方框 ②：内时钟信号源。

方框 ③：增量调制编、译码系统的时钟信号。

时钟速率有 64kHz、32kHz、16kHz 与 8kHz。目的是让学生通过选择不同的时钟速率观察增量调制编码与译码的波形，进一步加深对信道通频带带宽概念的理解。

方框 ④：CPLD 输出伪随机码作为系统的数据信息源。这样在进行通信系统综合实验时，既可以传输经过模/数变换后的音频信号（如语言信号、收录机播放的音乐信号或单一音频正弦波信号），又可以直接传输数据信息，让电路输出的伪随机序列信号波形的码元速率为 32kHz。

方框 ⑤：外加音频信号源。这在实验箱电路中是没有的，必须要由外部加入。它可以是语音信号,也可以是音乐信号,还可以是正弦波信号,信号输入通过开关 J106 选择。

方框 6：增量调制编码电路。它的作用是将输入的音频信号变成数字信号,它的时钟速率有 64kHz、32kHz、16kHz 及 8kHz 4 种方式供选择。它的编码方式是由 CPLD 控制器控制的。

方框 7：一个码型变换器。它的作用是把输入的绝对码变成相对码。这主要是为了在二相 PSK 调制方式中克服接收端的相位模糊而采用的。

方框 8：绝对移相键控(PSK)调制器。它根据输入数字基带信号的极性变化(1 码或 0 码)发送两个相位差为 π 的调相波,把发送基带信号的频谱搬移到适合于通信信道传输的频段,在实验中,频率为 1024kHz。

方框 9：1024kHz 载波发生器电路。

方框 10：一个二相 PSK 调相波信号接收放大电路。因为调相波信号经过信道传输后的波形变小了。因而在 PSK 解调器解调之前先把信号波形放大后再进行解调。

方框 11：载波提取锁相环电路。

前面已讲过,该电路实际是同相正交锁相环载波提取电路,它的特点是：在提取载波的同时,数字基带信号也已经解调出来。因此,解调出来的数字基带一方面送位同步恢复电路提取位同步信号,另一方面送到信码再生电路用位同步信号对该数字基带信码重新进行抽样判决。

方框 12：位同步恢复电路。它是从解调出的数字基带信号提取位同步信息,一方面作为再生码元信号抽样判决的定时基准,另一方面作为增量调制译码电路的抽样时钟信号。

方框 13：一个信码再生电路。它是用位同步信号作为再生时钟。对解调出的数字基带信号重新进行抽样判决。输出新的再生数字基带信码,一方面作为数据直接输出；另一方面作为增量调制译码电路的接收信码。

方框 14：码型反变换电路。它的功能是把判决再生的数字基带信码(相对码)恢复成绝对码序列。

方框 15：接收端增量调制系统译码电路。它是把数字化的音频信号恢复成模拟信号。它的取样速率有 64kHz、32kHz、16kHz,再生时钟 32kHz,可供选择。在进行该项综合实验时,只有选择再生时钟 32kHz,它的译码方式是由 CPLD 控制的。它恢复出的模拟信号要送入音频放大电路进行音频功率放大。

方框 16：音频功率放大电路。电路见实验五所示。它把增量调制译码电路输出的音频信号进行放大,去推动扬声器,把信息传送给收信者。

方框 17：再生信码或是码型反变换电路的输出再生信码。实质上是转换开关 K703 的 2 脚输出端。在此端可以测量输出数字信号波形。在测量点 TP711 处可观察到此波形。

至此,对通信系统综合实验按功能框图逐一进行了解释,使读者对组成一个较为完整的通信传输系统有了一定的了解。

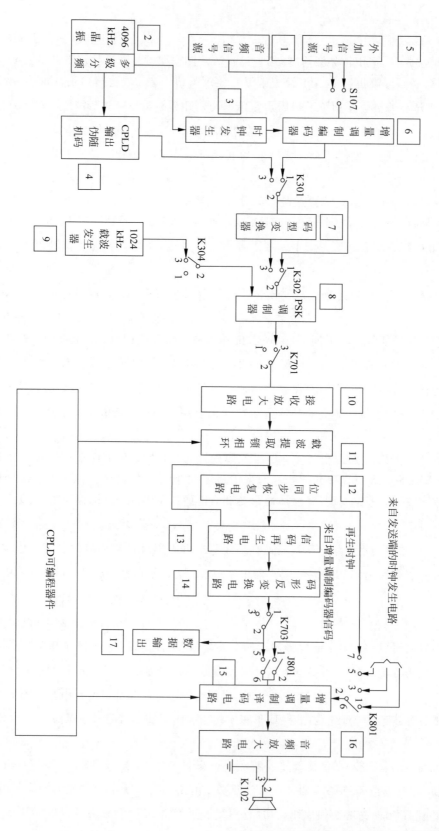

图 B.25　综合实验系统框图

3. 实验内容

1）用伪随机码序列做综合实验

先用伪随机码序列代替增量调制编码器输出的数字基带信号,逐一进行实验。最后从接收端中还原出伪随机码序列。

语音信号及通信系统原理综合实验按照图 B.25 进行连接。

2）单音频信号源做综合实验

断开伪随机码序列,用音频信号作信号源,送入增量调制系统编码电路的输入端,再逐一进行实验。最后从扬声器中还原出音频信号。

3）广播信号做综合实验

最后再用小话筒或录音机作信号源,逐一进行实验。观察实验的每一结果。

单台实验箱实现单工通信系统实验按照图 B.26 进行连接。

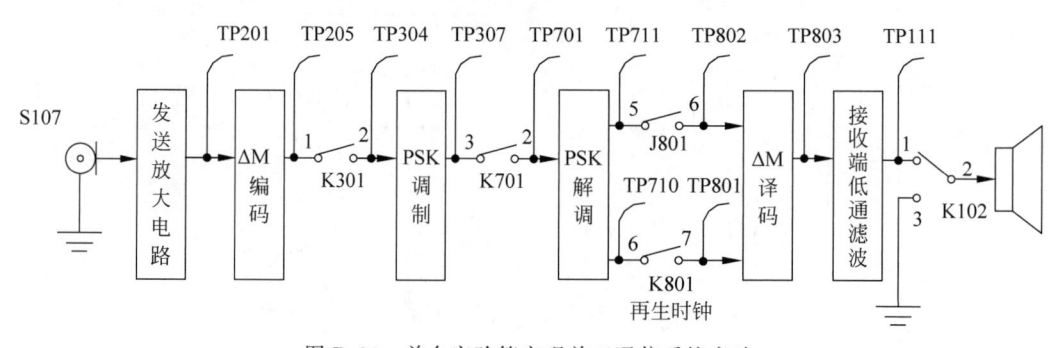

图 B.26 单台实验箱实现单工通信系统实验

4）两台实验箱实现双工编码综合实验

两台实验箱实现双工编码通信系统实验按照图 B.27 进行连接。

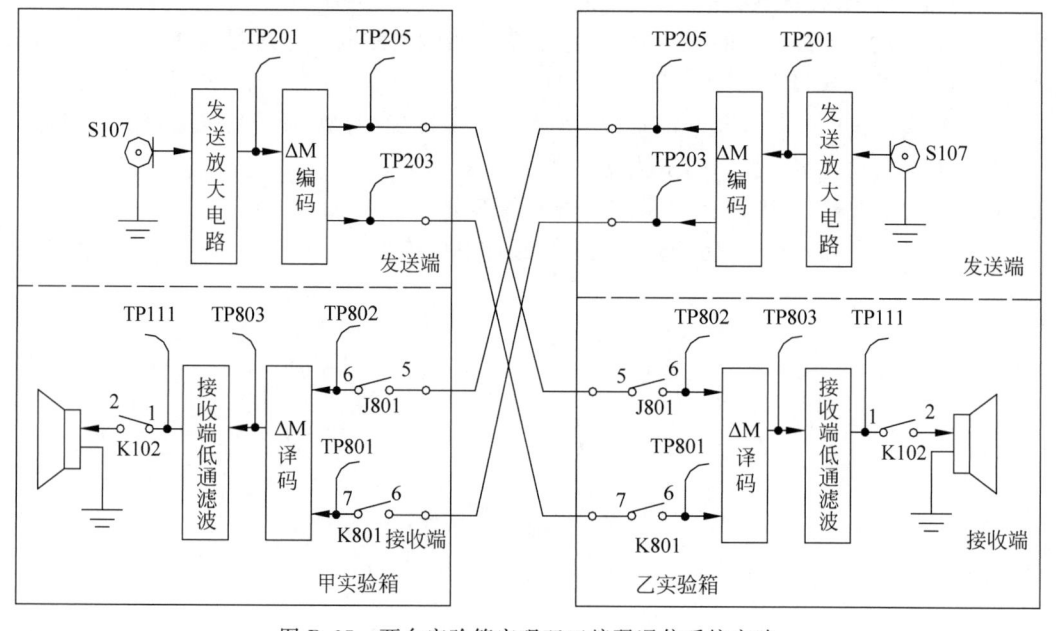

图 B.27 两台实验箱实现双工编码通信系统实验

用音频信号作信号源，送入甲实验箱的增量调制系统编码电路的输入端 S107 中，通过导线相连接到乙实验箱的增量调制系统译码电路的输入端 J801 的 5 脚与 6 脚相连，再逐一进行实验。最后从乙实验箱的扬声器中还原出音频信号。

5）两台实验箱实现双工调制解调综合实验

两台实验箱实现双工调制解调通信系统实验按照图 B.28 进行连接。

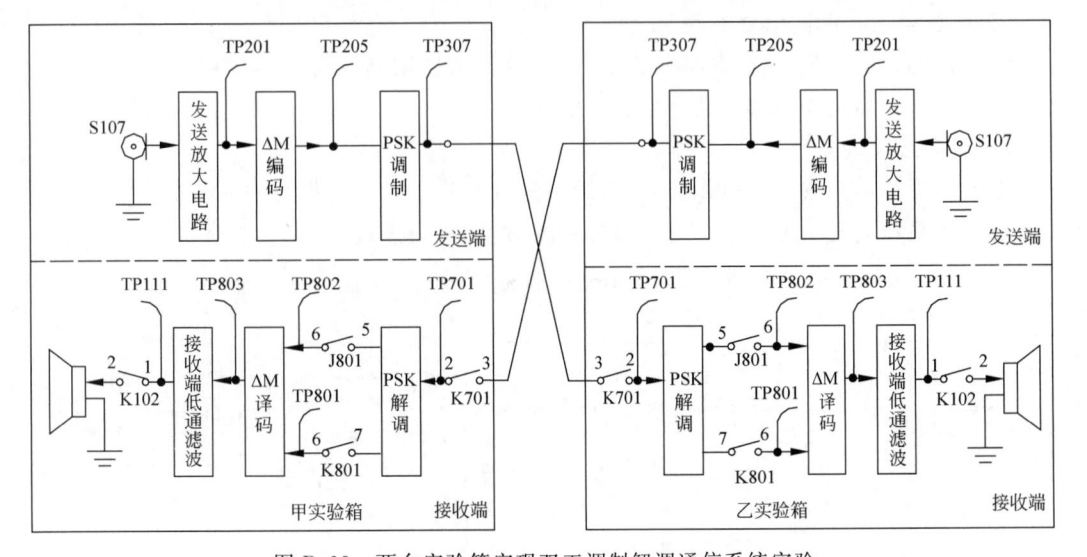

图 B.28　两台实验箱实现双工调制解调通信系统实验

用音频信号作信号源，送入甲实验箱的增量调制系统编码电路的输入端 S107 中，信号从 PSK 调制电路输出，通过导线相连接到乙实验箱的 PSK 解调电路的输入端 K701 的 2 脚与 3 脚相连，再逐一进行实验。最后从乙实验箱的扬声器中还原出音频信号。

4. 实验步骤及注意事项

（1）认真领会本次实验的指导思想，仔细分析实验电路的工作过程及原理。在动手之前做到实验系统基本概念清楚。综合实验目的明确。

（2）对实验箱中的各部分电路元器件所在位置看准确、清楚，具体步骤见以前的实验。

（3）根据实验内容中提到的步骤逐一进行。

（4）注意示波器、频率计、信号源的接地线要良好。

（5）每做一步都要小心，防止烧坏元器件或印制电路板。

误差函数表

$$\mathrm{erf}(x) = \frac{2}{\sqrt{\pi}} \int_0^x \exp(-y^2)\,\mathrm{d}y$$

x	$\mathrm{erf}(x)$	x	$\mathrm{erf}(x)$	x	$\mathrm{erf}(x)$	x	$\mathrm{erf}(x)$
0.00	0.000000	0.29	0.318283	0.75	0.711156	1.38	0.949016
0.01	0.011283	0.30	0.328627	0.76	0.717537	1.39	0.950673
0.02	0.022565	0.31	0.338908	0.77	0.723822	1.40	0.952285
0.03	0.033841	0.32	0.349126	0.95	0.820891	1.41	0.953852
0.04	0.045111	0.33	0.359279	0.96	0.825424	1.42	0.955376
0.05	0.056372	0.34	0.369365	0.97	0.829870	1.43	0.956857
0.06	0.067622	0.35	0.379382	0.98	0.834232	1.44	0.958297
0.07	0.078858	0.36	0.389330	0.99	0.838508	1.45	0.959695
0.08	0.090078	0.37	0.399206	1.00	0.842701	1.46	0.961054
0.09	0.101281	0.38	0.409009	1.01	0.846810	1.47	0.962373
0.10	0.112463	0.39	0.418739	1.02	0.850838	1.48	0.963654
0.11	0.123623	0.40	0.428392	1.03	0.854784	1.49	0.964898
0.12	0.134758	0.41	0.437969	1.04	0.858650	1.50	0.966105
0.13	0.145867	0.59	0.595936	1.05	0.862436	1.51	0.967277
0.14	0.156947	0.60	0.603856	1.06	0.866144	1.52	0.968413
0.15	0.167996	0.61	0.611681	1.07	0.869773	1.53	0.969516
0.16	0.179012	0.62	0.619411	1.08	0.873326	1.54	0.970536
0.17	0.189992	0.63	0.627046	1.26	0.925236	1.55	0.971623
0.18	0.200936	0.64	0.634586	1.27	0.927514	1.56	0.972628
0.19	0.211840	0.65	0.642029	1.28	0.929734	1.57	0.973603
0.20	0.222703	0.66	0.649377	1.29	0.931899	1.58	0.974547
0.21	0.233522	0.67	0.656628	1.30	0.934008	1.59	0.975462
0.22	0.244296	0.68	0.663782	1.31	0.936063	1.60	0.976348
0.23	0.255023	0.69	0.670840	1.32	0.938065	1.61	0.977207
0.24	0.265700	0.70	0.677801	1.33	0.940015	1.62	0.978038
0.25	0.276326	0.71	0.684666	1.34	0.941914	1.63	0.978843
0.26	0.286900	0.72	0.691433	1.35	0.943762	1.64	0.979622
0.27	0.297418	0.73	0.698104	1.36	0.945561	1.65	0.980376
0.28	0.307880	0.74	0.704678	1.37	0.947312	1.66	0.981105

续表

x	erf(x)	x	erf(x)	x	erf(x)	x	erf(x)
1.67	0.981810	2.09	0.996880	2.51	0.999614	2.93	0.999965
1.68	0.982493	2.10	0.997021	2.52	0.999635	2.94	0.999968
1.69	0.983531	2.11	0.997155	2.53	0.999654	2.95	0.999970
1.70	0.983790	2.12	0.997284	2.54	0.999672	2.96	0.999972
1.71	0.984407	2.13	0.997407	2.55	0.999689	2.97	0.999973
1.72	0.985003	2.14	0.997525	2.56	0.999706	2.98	0.999975
1.73	0.985578	2.15	0.997639	2.57	0.999722	2.99	0.999977
1.74	0.986135	2.16	0.997741	2.58	0.999736	3.00	0.99997791
1.75	0.986672	2.17	0.997851	2.59	0.999751	3.01	0.99997926
1.76	0.987190	2.18	0.997957	2.60	0.999764	3.02	0.99998053
1.77	0.987691	2.19	0.998046	2.61	0.999777	3.03	0.99998173
1.78	0.988174	2.20	0.998137	2.62	0.999789	3.04	0.99998286
1.79	0.988164	2.21	0.998224	2.63	0.999800	3.05	0.99998392
1.80	0.989091	2.22	0.998308	2.64	0.999811	3.06	0.99998492
1.81	0.989525	2.23	0.998388	2.65	0.999822	3.07	0.99998586
1.82	0.989943	2.24	0.998464	2.66	0.999831	3.08	0.99999674
1.83	0.990347	2.25	0.998537	2.67	0.999841	3.09	0.99998757
1.84	0.990736	2.26	0.998607	2.68	0.999849	3.10	0.99998835
1.85	0.991111	2.27	0.998674	2.69	0.999858	3.11	0.99998908
1.86	0.991472	2.28	0.998738	2.70	0.999866	3.12	0.99998977
1.87	0.991821	2.29	0.998799	2.71	0.999873	3.13	0.99999042
1.88	0.992156	2.30	0.998857	2.72	0.999880	3.14	0.99999108
1.89	0.992479	2.31	0.998912	2.73	0.999887	3.15	0.99999160
1.90	0.992790	2.32	0.998966	2.74	0.999893	3.16	0.99999214
1.91	0.993090	2.33	0.999016	2.75	0.999899	3.17	0.99999264
1.92	0.993378	2.34	0.999065	2.76	0.999905	3.18	0.99999311
1.93	0.993656	2.35	0.999111	2.77	0.999910	3.19	0.99999356
1.94	0.993923	2.36	0.999155	2.78	0.999916	3.20	0.99999397
1.95	0.994179	2.37	0.999197	2.79	0.999920	3.21	0.99999436
1.96	0.994426	2.38	0.999237	2.80	0.999925	3.22	0.99999478
1.97	0.994664	2.39	0.999275	2.81	0.999929	3.23	0.99999507
1.98	0.994892	2.40	0.999311	2.82	0.999933	3.24	0.99999540
1.99	0.995111	2.41	0.999346	2.83	0.999937	3.25	0.99999570
2.00	0.995322	2.42	0.999379	2.84	0.999941	3.26	0.99999598
2.01	0.995525	2.43	0.999411	2.85	0.999944	3.27	0.99999624
2.02	0.995719	2.44	0.999441	2.86	0.999948	3.28	0.99999649
2.03	0.995906	2.45	0.999469	2.87	0.999951	3.29	0.99999672
2.04	0.996086	2.46	0.999497	2.88	0.999954	3.30	0.99999694
2.05	0.996258	2.47	0.999523	2.89	0.999956	3.31	0.99999715
2.06	0.996423	2.48	0.999547	2.90	0.999959	3.32	0.99999734
2.07	0.996582	2.49	0.999571	2.91	0.999961	3.33	0.99999751
2.08	0.996734	2.50	0.999593	2.92	0.999964	3.34	0.99999768

续表

x	erf(x)	x	erf(x)	x	erf(x)	x	erf(x)
3.35	0.999997838	3.52	0.999999358	3.69	0.999999820	3.86	0.999999952
3.36	0.999997983	3.53	0.999999403	3.70	0.999999833	3.87	0.999999956
3.37	0.999998120	3.54	0.999999445	3.71	0.999999845	3.88	0.999999959
3.38	0.999998247	3.55	0.999999485	3.72	0.999999857	3.89	0.999999962
3.39	0.999998367	3.56	0.999999521	3.73	0.999999867	3.90	0.999999965
3.40	0.999998478	3.57	0.999999555	3.74	0.999999877	3.91	0.999999968
3.41	0.999998583	3.58	0.999999587	3.75	0.999999886	3.92	0.999999970
3.42	0.999998679	3.59	0.999999617	3.76	0.999999895	3.93	0.999999973
3.43	0.999998770	3.60	0.999999644	3.77	0.999999903	3.94	0.999999975
3.44	0.999998855	3.61	0.999999670	3.78	0.999999910	3.95	0.999999977
3.45	0.999998934	3.62	0.999999694	3.79	0.999999917	3.96	0.999999979
3.46	0.99999008	3.63	0.999999716	3.80	0.999999923	3.97	0.999999980
3.47	0.999999077	3.64	0.999999736	3.81	0.999999929	3.98	0.999999982
3.48	0.999999141	3.65	0.999999756	3.82	0.999999934	3.99	0.999999983
3.49	0.999999201	3.66	0.999999773	3.83	0.999999939		
3.50	0.999999257	3.67	0.999999790	3.84	0.999999944		
3.51	0.999999309	3.68	0.999999805	3.85	0.999999948		

参 考 文 献

［1］ Candès E J，Romberg J，Tao T. Robust uncertainty principles：Exact signal reconstruction from highly incomplete frequency information［J］. IEEE Trans. Inf. Theory，2006，52(2)：489-509.

［2］ Foucart S，Rauhut H. A Mathematical Introduction to Compressive Sensing［M］. New York：Springer Science & Business Media，2013.

［3］ https：//www. codeproject. com/Articles/852910/Compressed-Sensing-Intro-Tutorial-w-MATLAB.

［4］ Rani M，Dhok S B，Deshmukh R B. A Systematic Review of Compressive Sensing：Concepts，Implementations and Applications［J］. IEEE Access. 2018，6：4875-4894.

［5］ Maria-Gabriella，Benedetto D，Giancola G. 超宽带无线电基础［M］. 葛利嘉，朱林，袁晓芳，等译. 北京：电子工业出版社，2005.

［6］ Wang B J，Xu H，Yang P. Target Detection and Ranging through Lossy Media using Chaotic Radar［J］. Entropy 2015，17：2082-2093.

［7］ https：//www. nutaq. com/blog/introduction-mimo-wireless-communications.

［8］ https：//cdn. rohdeschwarz. com/pws/dl _ downloads/dl _ application/application _ notes/1ma142/1MA142_0e_introduction_to_MIMO. pdf.

［9］ Wu X. Advanced multiple input multiple output（MIMO）SAR algorithm for high-resolution 3D reconstruction imaging［C］. 2017 XXXIInd General Assembly and Scientific Symposium of the International Union of Radio Science（URSI GASS），19-26 Aug. Montreal，QC，Canada，2017：1-4.

［10］ 樊昌信，曹丽娜. 通信原理（精编本）［M］. 北京：国防工业出版社，2008.

［11］ 樊昌信. Principles of Communications［M］. 北京：电子工业出版社，2010.

［12］ Haykin S. Communication Systems［M］. New York：John Wiley & Sons，2001.

［13］ 王俊峰，孙江峰. 通信原理 MATLAB 仿真教程［M］. 北京：人民邮电出版社，2010.